# YOUR BRAIN ON EXERCISE

# YOUR BRAIN
# ON EXERCISE

**GARY L. WENK, PHD**

OXFORD
UNIVERSITY PRESS

# OXFORD
### UNIVERSITY PRESS

Oxford University Press is a department of the University of Oxford. It furthers
the University's objective of excellence in research, scholarship, and education
by publishing worldwide. Oxford is a registered trade mark of Oxford University
Press in the UK and certain other countries.

Published in the United States of America by Oxford University Press
198 Madison Avenue, New York, NY 10016, United States of America.

Library of Congress Cataloging-in-Publication Data
Names: Wenk, Gary L, author.
Title: Your brain on exercise / Gary L. Wenk, PhD.
Description: New York, NY : Oxford University Press, [2021] |
Includes bibliographical references and index.
Identifiers: LCCN 2020025000 (print) | LCCN 2020025001 (ebook) |
ISBN 9780190051044 (hardback) | ISBN 9780190051068 (epub)
Subjects: LCSH: Brain—Physiology. | Exercise—Health aspects.
Classification: LCC QP376 .W465 2021 (print) | LCC QP376 (ebook) |
DDC 612.8/2—dc23
LC record available at https://lccn.loc.gov/2020025000
LC ebook record available at https://lccn.loc.gov/2020025001

1 3 5 7 9 8 6 4 2

Printed by Sheridan Books, Inc., United States of America

*This book is dedicated to my amazing wife, who insisted that every day I close my computer and join her for a long walk. Her wisdom gave birth to the core of my story. This book is dedicated to Jane.*

# CONTENTS

# ACKNOWLEDGMENT

I cannot express enough thanks to Joan Bossert, my wonderful editor at Oxford University Press. I offer my sincere appreciation for her suggestion to take on this fascinating and challenging topic and for her inspired guidance along the way.

# INTRODUCTION

When a new drug is discovered, such as Prozac, or an old drug is rediscovered, such as marijuana, claims that the long-sought-after panacea has been finally discovered reverberate through social media; books are written extolling the amazing health benefits that everyone can now easily access. Whatever your actual illness or mental condition, for only a few dollars, this new drug is your answer. In my long career studying the effects of drugs and nutrients on the brain I have witnessed the ephemeral excitement quickly fade after closer examination. I depend on my colleagues in other research laboratories, as they depend on me, to report the truth and tamp down the enthusiasm created by false claims. This correction process always happens. Unfortunately, sometimes it can take a while.

Exercise has become one of these new "drugs." The claims about the benefits of exercise are numerous and sometimes outrageous. Some claims are true and are supported by available scientific evidence, while others are not. Decades of excellent research have demonstrated that regular modest levels of exercise improve heart and lung function and may relieve joint pain. Regular daily exercise will help your body to regulate blood sugar levels and reduce inflammation; many of these benefits are a consequence of reducing the amount of body fat you carry around. Your body clearly benefits in many ways from regular exercising.

But does your brain benefit as well? The answer depends on how you frame the question. Publications in the scientific and popular literature provide a confusing blend of information and misinformation about the potential benefits of exercise for the brain. In my opinion, our current state of knowledge about the benefits and risks of exercise parallels the state of our knowledge about the benefits and risks of drugs that act on the brain. There is very little reliable information to make definitive statements that apply to everyone under all possible conditions of health and illness.

Consider one popular drug, marijuana. Depending on your particular political, religious, or cultural bias it is possible to find evidence in the scientific literature that can support a strong case both for and against the use of marijuana. It all depends on how you ask the question and how well the scientific inquiry was conducted. As I prepared this book, I discovered that the popular literature contains many claims about the benefits of exercise on the brain that the available scientific evidence does not entirely support. Many times, original scientific publications were quoted by authors who lack the necessary training to understand the article and who then misinterpreted the findings to fit into their preconceived narrative about the benefits of exercise. Needless to say, similar misrepresentations exist regarding the benefits and risks of medical marijuana.

In order to understand the effects of exercise on the human brain, it is necessary to understand how your brain functions and how your muscles communicate with your brain to influence those functions. The misrepresentations in the popular literature persist and are rarely corrected because the actual story is quite complicated and there is still a lot to be discovered about the effects of exercise on brain function. Whenever there is a gap in our current knowledge, myths rush in to fill the empty spaces until they are swept out by new discoveries and a more accurate understanding. For these reasons some myths about exercise have been much harder to dislodge than others.

One thing is certain: Regular exercise benefits the body. Does regular exercise positively affect brain function? Does our thinking become faster because we exercise? Does running a marathon make us smarter? My goal in the following chapters is to provide a realistic perspective on what benefits your brain should expect to achieve from exercise. When you consume any medicine, you have expectations about the benefits that you should obtain and also recognize that there are consequences; these are called side effects. For example, aspirin relieves skeletal and muscular pain but also irritates the lining of the stomach and intestines and causes increased intestinal bleeding. In addition, taking more of a drug often (but not always) provides more benefit but also leads to more severe side effects. We accept the risks in order to obtain the benefits. As you will learn in the following chapters, the same overall principle holds true for exercise; a little is good, but too much can be harmful, particularly for the brain. The following chapters will discuss why this is so.

In order for exercise to influence the brain, the muscles involved must somehow communicate with it. Long ago, it was thought that active muscles communicated with the brain via electrical impulses from sensory nerves originating within muscles. This method of communication with the brain is now considered highly unlikely. When the spinal cord is severed, thus preventing both afferent and efferent electrical flow into and out of the brain, contraction of paralyzed muscles by direct electrical stimulation generated the same beneficial physiological changes in the brain as seen in active muscles of normal individuals. Therefore, the current understanding is that actively contracting skeletal muscles communicate with the brain, as well as many other organs, by releasing chemical messengers into the blood. During the past few years, many muscle-derived chemical messengers have been discovered—so many, in fact, that your muscles might be more accurately viewed as one of your endocrine glands, similar to your adrenal or thyroid glands.

The actions of these chemical messengers on your brain, and whether these actions are direct or indirect, beneficial or not, is the focus of this book. Initially, I expected that the story would be clear and the published literature would be consistent and demonstrate a significant beneficial effect of exercise on brain function. I was mistaken. My discoveries led me to the realization that Friedrich Wilhelm Nietzsche was right when he stated that "the devil is in the details." The familiar phrase warns that at first glance something might seem simple and easily understood, however, with closer examination the details of a matter are become its most problematic aspect. When I examined the most recent scientific literature, I discovered that the truth is more nuanced and far more interesting than I initially expected. I hope that you enjoy this journey of discovery as much as I did.

# Part I

# THE HIGH COST OF EXERCISE

# 1

# LEARNING FROM *THE BIGGEST LOSER*

The singer Tennessee Ernie Ford was correct: It is important to know your audience, but it also important that you know your author so that you understand my approach to this topic. I am a college professor and research scientist. For the past 40-plus years I have been teaching psychology and biology undergraduates, first-year medical students, neurology residents, and psychiatrists about the workings of the brain. I am a neurochemist and pharmacologist by training. My laboratory research focused on how food and drugs influence how we think and feel and how well we age. Thus, my emphasis in this book is on the brain and how its health and function are influenced by exercise.

My young undergraduate students are endlessly fascinated by how their brain works. You might predict that this fascination is motivated by their natural curiosity and desire to advance their education. You would be wrong. The most common question that I am asked during my office hours is "Hey, Doc, I ate (or drank) this last night. What do you think it might do to me?" while holding out an empty vial or package.

After hearing this question from my 19-year-olds for so many years, I have come to the conclusion that there is only one explanation for their behavior: They feel immortal. They believe that because they are young, they are immune to the dangers of life. As a brain scientist, I blame these feelings of immortality on the fact that their frontal lobes are not fully working because their brains have not yet completed the maturation process. Neuroscientists discovered many years ago that the parts of human brain that evolved most recently, and thus give rise to the sophisticated mental abilities that we usually associate with adult behavior, are the parts that mature last. If you are a parent, that last statement did not surprise you. The frontal lobes, the part of the brain that contributes most to our unique personalities and allows us to anticipate the consequences of our imprudent actions, matures last. Essentially, the frontal lobes of my students should tell them that it's a bad idea to drink alcohol and drive, or to ignore the consequences on their health of eating cheeseburgers and pizza every day, or the costs of binge drinking every Thursday evening. Usually, when their frontal lobes finally complete the maturation process, they stop doing stupid things. Most importantly, they should stop acting as though they are immortal. Females do not have fully mature brains until they are about 25 years old; unfortunately, the males must wait until they are about 30 years old to enjoy the benefits of a fully matured brain. Every semester I like to remind my 19-year-old female students that dating a guy who is also 19 years old is roughly equivalent to dating a boy with a 16-year-old brain. At this point, the females in the classroom are typically nodding their heads in recognition that what I just said is painfully familiar.

Year after year, my 19-year-olds offer testimonials in class, in front of about 200 of their peers (they never hesitate to over-share in a game of pharmaceutical one-upmanship), detailing the evidence of their immature brain structure. One young female student told this story in class one morning: She said that she had met a woman in a bar the previous night who claimed to be a "nurse" (an immature brain also makes one rather gullible) who offered to give her some pills. She admitted that she had no idea what these pills contained but was promised by her new friend that it would be a fun experience. She accompanied the "nurse" to her apartment that was nearby. Her experience included hallucinations, heart palpitations, and a fever, consistent with having consumed the popular street drug Ecstasy. What I found astonishing was that this young woman did not seem at all disturbed by the fact that she had willingly allowed a total stranger to place her health at risk. She told the class that when she woke in the morning she got dressed (she did not remember getting undressed) and came to class. For reasons that I will never understand fully, she was eager to share her adventures with the entire class.

The male brain completes its final stages of maturation much later than female brains. Thus, my 19-year-old male students tend to make decisions that are equally reckless, but for a much longer period of time. One of my male students described the following experience in class. During the previous weekend he became bored and decided that it might be fun to ingest a packet of instant coffee, right out of the box. He popped an entire packet (paper included) into his mouth and swallowed it. Apparently, he enjoyed the rush so much that he decided to finish off the entire box of 32 packets! Three days later, he stopped having explosive diarrhea and finally fell asleep, completely exhausted. You can now fully appreciate why car insurance rates go down for women at age 25 years and for men at age 30 years; it has everything to do with the maturation of the frontal lobes.

I once thought that the widespread access to all knowledge on the internet would allow my students to inform themselves of the risks of certain behaviors before they could get themselves into serious trouble. Sadly, this has never been the case; the internet tends to give them a false sense of security. For example, one young man announced to the class that he had discovered an opium poppy growing on his property during the past weekend. He checked on the internet to confirm the identity of the plant. I asked him what he did with it. I assumed that after checking the internet he would discover that he should handle this plant very carefully and certainly not eat it. I was wrong; he proudly announced to the class that he ate it.

"The entire plant!?" I said.

"Yes, of course," he said.

I then asked the most obvious question that was certainly on the minds of everyone else in the room that day: "What happened?"

He said, "A couple hours later my girlfriend found me unconscious in the yard and called 911. It was an amazing experience."

Not for a second did he believe that his life was at risk by his reckless behavior. He felt immortal.

Decades of unbelievable stories by these immortal young men and women led me to write a book called *Your Brain on Food* (Oxford University Press) that I use as the textbook for my *Introduction to Psychopharmacology* course. The main point of the book is that anything that you consume can affect your brain function, how you think, how you feel, and will influence how fast your brain ages. I discovered that my students displayed caution when using over-the-counter or prescription drugs but often demonstrated a cavalier, and occasionally irresponsible, indifference to the effects of what they ate or drank. After all, if it's "food" it must be safe. I wrote the book to help them understand why they should be cautious about anything that they consume, whether it is a food or drug or something in between, such as chocolate or coffee.

Obsessions with selected drugs are not the only topics they discuss during class or office hours. Many of my students are also obsessed with some form of exercise, such as team or individual sports or one of the many popular extreme sports such as 72-hour jogging, paraskiing, or volcano surfing (it's actually a lot of fun, although it rips your clothes to shreds). Each year, many of my students talk about their daily yearning to run or swim; these are the obligatory athletes. They are driven to run 10 to 20 miles or swim four hours every day, no matter how inconvenient to their daily schedule or how much physical pain it produces. This addiction occurs primarily with aerobic exercises rather than anaerobic ones. I will discuss why this is so in a later chapter. Frequently, my obligatory athletes come to class and promptly fall asleep from exhaustion. Needless to say, their classroom performance suffers due to their exercise addiction. They are as addicted to extreme exercising as a drug addict is to heroin. Later I will explain how identical brain regions are responsible for both addictions.

Many of my obligatory athletes defend their addiction by claiming that daily exercise is the only way they can lose weight and remain healthy. Heroin addicts usually have explanations that also sound rational. Is an exercise addiction really as harmful to your health as a drug addiction? In many ways, yes, if it is taken to excessive levels. One of the main problems with an exercise addiction is that it requires that the obligatory athlete consume 3,000 to 4,000 calories every day. My obligatory athletes are constantly eating or thinking about eating.

Consuming so many calories every day has a negative effect on overall health. Why? Due to the increased requirement for additional oxygen to metabolize all of that food, high-calorie diets accelerate the aging process. The relationship is linear: The more you eat, the faster you age and the sooner you die. Ironically, their initial desire to become healthier by jogging to lose weight addicted them to the extreme exercising routine that required them to consume thousands of extra calories that accelerated their aging. The biochemical processes inside all living multicellular creatures that link eating with movement and tissue aging were forged billions of years ago and cannot be circumvented. You are probably well aware that what you eat every day influences your health in the short term; however, how much you eat every day also influences your health over the long term and determines how fast your body ages. Although I discuss this topic in class every year, the lesson never seems to stick.

Unfortunately, my students' obsessions also include adherence to one of the many popular diets that are either unhealthy or have absolutely no scientific basis, such as the paleo diet, the Atkins diet, "The Zone," the Dukan diet, the apple cider vinegar diet, the coconut oil diet, the low-carb/high-protein diet, the high-carb/low-fat diet, the cotton ball diet, or the misleadingly named and most scientifically goofy ketogenic diet. These diets frequently produce harmful effects on the kidneys, heart, and liver that undermine any potential health benefits that might be obtained from exercising. I do not understand why my students expend so much effort to exercise and then undermine their efforts by following the foolish advice associated with these fad diets.

One of the most popular fad diets lately is going gluten-free. Many of my students have decided to become gluten-free for absolutely no medical reason. Is eating gluten always bad for everyone? Absolutely not. If you are not gluten-sensitive, then avoiding gluten is a bad idea according to the results of a large research study involving over 15,000 participants who were followed for 30 years. The American College of Cardiology now strongly recommends against the adoption of gluten-free diets for people without a medical necessity. Thus, unless you suffer with celiac disease, going gluten-free is not a good dietary decision. Gluten-free diets, like so many other fad diets of the past and present, have been promoted by uninformed or willfully ignorant nutritional prophets (too often by phycians, who should know better) using dietary scare tactics to scam a few dollars from a misinformed public.

The interesting thing about diets and exercising is that we are always being instructed to get the right amount of both. What precisely does this mean? Usually, the goal is to avoid becoming obese so that we can live a long, healthy, and active life. Why is being obese so unhealthy? The culprit is the same one that explains why it is so unhealthy to consume 4,000 calories every day: oxygen, and your requirement for a constant supply of it in order to metabolize your food.

Thanks to the impact of social media, the fashion industry, government health agencies, and high school health classes, everyone is aware that carrying around too many extra pounds of fat is unwise. Obesity reproduces, and also accelerates, many of the same metabolic conditions that underlie the normal aging process. Decades of research have demonstrated that obesity increases the risk of diabetes, atherosclerosis, and especially cancer and, with regard to its effects on the brain, increases the likelihood of becoming depressed and then makes it harder for antidepressant therapies to work. For the past few decades I have witnessed first-hand the negative consequences of obesity on the quality of the lives of many of my students. My obese students are often shunned by others and perform more poorly in class than students who are thinner. Notice that the relationship is between amount of body fat and academic performance, not between exercise and academic performance. As will become clear later, exercise does not significantly enhance cognitive performance; however, body fat does significantly impair cognitive performance.

Is there a biological relationship between being obese and not doing well in school? Yes, there is. Does excess body fat make you stupid? Yes, it does. Numerous epidemiological and clinical studies have clearly demonstrated that obesity is a definitive risk factor for cognitive deficits and impaired memory in young and old humans. One recent epidemiological study concluded that by the year 2050 the percentage of overweight or obese Americans will increase to over 50%. This statistic is alarming because of the severe consequences on the body and brain of being obese. Obesity also increases the incidence of numerous diseases that will place great demands on the nation's medical bill, particularly Medicaid and Medicare, as the general population gets older. Why?

Fat is toxic to your body and brain. Many laboratories around the world, including my own, have documented the mechanisms that underlie how excessive body fat impairs brain function. Scientists have focused on the obesity-induced degenerative changes that occur within a brain structure called the *hippocampus*. The normal function of the hippocampus is critical for the formation of new memories. This brain structure will be the focus of many later discussions about the effects of exercise on the brain. Recent studies have reported specific abnormalities in the general health of cells within the hippocampus that underlie the memory impairments related to obesity.

All of my students were told in elementary school that they should eat right and exercise so that they perform at their personal best in school or at work. What is the optimal balance of food and exercise in order to obtain a healthy body? My impression is that my students were not given enough information to make this judgement for themselves, because year after year I get this question: How much should I exercise every day to become healthier? Unfortunately, that is the wrong question. The answer has nothing to do with exercising; rather, I tell them that how little they eat every day is far more important than how much they exercise every day.

I am not the only person who preaches this sermon. The trainer for the popular NBC television show "The Biggest Loser" used to think that more exercise was all that was necessary in order to lose weight and get healthy. After many years of helping severely obese people lose weight, however, Bob Harper concluded that exercise is not the key; diet is far more important than exercise. Not only is Harper helping his clients to feel better and achieve their personal goals, he is also helping them to live longer, healthier lives. His clients are now instructed to consume fewer calories every day in order to reduce their level of excess body fat.

Fat accelerates aging and increases your risk of dying. How? Fat cells produce inflammation. Lots of it. What is the best way to reduce inflammation in your body? In order to answer this question, scientists investigated whether diet or exercise more effectively reduced the levels of inflammation in overweight or obese women. After 12 months of investigation the scientists concluded that the greatest weight loss and most significant reduction in the level of inflammatory protein reduction came *only* from dieting. The women who participated in an exercise-only program showed no reduction in inflammatory proteins. The reasons for this lack of change in the level of inflammation following exercise will become apparent in later chapters. Essentially, unless you are a marathon runner, a triathlete, or an Olympic-level swimmer, the activity of your musculature is not a big player in calorie consumption or body fat reduction.

Although the results of these studies on humans are interesting, sometimes it is possible to get more detailed data from studies of nonhuman primates. Fortunately, nonhuman primates share critical features of our physiology. In order to investigate the benefits of caloric restriction, a large group of monkeys, ranging from middle-aged to quite elderly, were fed only 70% of their free-feeding diet for about 15 years. For someone eating a 2,000-calorie-per-day diet, this would be about 600 fewer calories per day. As a result of eating just 30% fewer calories, the brains of the monkeys on the restricted diet aged significantly more slowly, and the monkeys developed far fewer age-related diseases, had virtually no indication of diabetes and almost no age-related muscle atrophy, and lived much longer. Most importantly, and consistent with Bob Harper's conclusion, these monkeys did not exercise the weight off; they simply consumed fewer calories every day. This was a rather shocking discovery: What is the value of exercising, if not to lose weight? As will become clear in the following chapters, moderate daily exercising does offer real health benefits for your body. But that's not the question being asked in this book: Does your brain benefit as well?

I discovered that the answer is more complicated and more interesting than I had expected. Some of our assumptions about the benefits of exercise are simply not true. Although a little exercise is good, too much can produce harm. Furthermore, the type of exercising directly influences the benefits we obtain. In many ways, the effects of exercise on the body parallel those of the effects of food and drugs that I described in my book *Your Brain on Food*. The puzzling discoveries I made while reviewing the available scientific literature led me to a conclusion that became the title of the next chapter.

# 2

# THE DEVIL IS IN THE DETAILS

Movement has always been a required part of survival; thus, evolution has finetuned the enzymatic machinery that generates the energy required for physical activity. The environmental forces that drove these evolutionary processes never anticipated that in the late 20th century a species would develop a culture that effectively disconnected the ancient link between physical activity and survival. You can remain physically inactive and still survive in our culture. You can earn a living while sitting in your pajamas at your computer. Fully prepared, healthy meals can now be delivered to your door; no need to go out foraging and risk injury or death. Consequently, today, physical inactivity has emerged as a major risk factor for your brain health and survival. One of the best and most comprehensive studies investigating the relationship between physical fitness and mortality found that both men and women lived significantly longer by exercising regularly. Avoiding a sedentary lifestyle delays death primarily due to lower rates of cardiovascular disease and cancer.

Nevertheless, should being too sedentary be considered equivalent to smoking, as some authors have claimed? Not really. Smoking is considerably worse than being sedentary; however, if given the choice, you should do neither. Indeed, as will become clear later, obesity, not inactivity, should be considered equivalent to smoking. Obesity impairs cognition and increases the risk for some psychiatric disorders and dementias and is responsible for far more hospitalizations and deaths every year in the United States than tobacco smoking. Worse, obesity is an intergenerational problem; if the mother and father are obese, they significantly predispose their off-spring to poor cognitive abilities.

But are we all really that inactive? In spite of warnings on the internet that are primarily intended to motivate us to buy another treadmill or a colorful pair of yoga pants, most of us probably achieve a low level of modest exercise every day that benefits our overall health. If we are to believe the data from people using devices that measure their daily number of steps, a healthy person usually takes about 6,000 to 7,000 steps per day just completing their normal daily tasks. A 30-minute stroll after dinner could add another 3,000 steps—if anyone bothered to do so. Overall, such a routine could provide the essential level of exercise that today's health experts recommend for good cardiovascular health. However, is this much exercise per week sufficient to keep your brain healthy and allow it to perform its two most basic functions? The answer is a highly qualified *yes*, especially if you restrict your calorie intake, but more detailed research needs to be performed in order to be certain.

Brains evolved to perform two basic functions that usually involve muscle activity: your personal survival and the procreation of your species. This is true for all known animals, even those with the most simplistic nervous systems. Your brain utilizes information gathered by your sensory systems, considers its options, and then instructs your muscles to contract in order to obtain food, find receptive mates for reproduction, or avoid predators. Thus, your brain achieves its primary goals via control of your muscles. However, beyond the obvious benefit to survival and procreation, does your brain benefit from the simple act of having your muscles contract? Is exercise beneficial to the brain directly, or only indirectly via its effects on the body?

In order to answer these questions, I first need to define exercise. Exercise is a subset of all possible physical movements. For most of us who are not professional athletes, exercise is a planned, socially structured, and repeated behavior performed with the intent to maintain or improve some aspect of physical fitness. Given this definition, it is unlikely that other animals participate in exercise for the same reasons offered by human athletes. Exercise is a special form of movement that is not related to finding food or a receptive mate (ignoring the predatory behaviors of males at fitness clubs).

A lack of regular exercise, sedentary lifestyles that do not require much activity, and high-calorie diets together synergistically impair body and brain health. How? The answer is simple: These behaviors lead to increased numbers of fat cells, and fat cells produce widespread oxidative stress and inflammation. Inflammation is very harmful because it predisposes us to disease, injury, cancer, and early death.

### The inflammation inferno

*Inflammation* refers to high concentrations of a group of proteins that originally evolved to protect us from invading organisms and poisons in the environment. Today, due to our high-calorie diet and sedentary culture, humans have almost entirely flipped the function of these inflammatory proteins from one of protection to one of harm. Thanks to our tendency to eat too much and move too little, the additional fat that we carry around inside our body leads to the release of ever-increasing levels of inflammatory proteins that have toxic effects.

How can you best protect yourself from these toxic inflammatory proteins? Numerous recent studies have concluded that diet is far more effective than exercise for reducing the levels of inflammatory proteins in obese humans. Study after study has demonstrated that the greatest weight loss and the most significant reduction in the level of inflammatory proteins come from reducing your total caloric intake. These results are very important; I will return to the consequences of inflammation when I discuss how exercise affects the brain.

For those people who think that they can eat whatever they wish and then simply exercise away the calories (I am talking primarily to young males here), the news is not good. In some ways, our bodies negatively adapt to daily physical activity. Our bodies work against our attempts to exercise off those extra pounds via processes that evolved millions of years ago and were originally intended to allow our bodies to maintain their total energy expenditure within a safe narrow range when food was scarce. The scarcity of food for our ancestors drove many of the adaptive physiological changes that form the basis of the principles outlined in the following chapters. These adaptive mechanisms reduce the amount of energy used for movement in order to conserve it for other body functions. Thus, due to this adaption, it ultimately becomes more difficult to lose weight by only exercising.

It gets worse! Too much exercising for too long can suppress immune function and reduce reproductive success. Essentially, for most of us, the activity of our musculature is never going to consume the bulk of our calories. Scientists speculate that as we evolved our big, fast-thinking, energy-demanding brain our ancestors lost the ability to metabolize a substantial number of calories via physical activity. Our brain took priority over our muscles.

### Your intestinal factory

The nutrients from your last meal are currently being absorbed by your small intestines (mostly) and will soon become available to all of the cells of your body. The entire gastrointestinal system, including stomach, large and small intestines, and their related organs, such as the pancreas and liver, utilize nearly 70% of the energy you just consumed in order to make the remaining 30% available to the rest of your body.

Your brain uses about 20% of the remaining available energy, and your other organs that allow you to reproduce and move around your environment (including your muscles and bones) utilize what is left over. As you can see, very little energy is left over for other tasks in the body.

These percentages give you some idea of the priorities—thinking about finding food and avoiding predators, finding and reproducing with other humans, and finally, the physical activity required to accomplish these goals—that billions of years of evolution have set for your body to achieve. About 200,000 years ago the predecessors of our species traded a big brain and enhanced survival abilities for reduced reproductive success and fewer, smaller muscles. The devil is always in the details: This tradeoff underlies why humans do not give birth to litters and why it is so hard to exercise away those extra pounds. However, exercise does provide benefits, as long as it is performed *in moderation*. Your brain will not tolerate being robbed of its energy for too long; there will be consequences.

### All things in moderation

Overall, exercise can improve health; however, maximal benefits are achieved when moderate levels of physical activity are combined with reduced calorie intake. The principle of moderation—not too much and not too little—applies to both exercising and eating. Why? The answer is related to how the cells of your body convert the food you eat into the energy required to contract your muscles. This process also underlies why either eating or exercising too much is harmful to your body and brain. In order to understand why, we need to travel back in time.

# 3

# A LONG TIME AGO

Our search to discover the origin of the mechanisms that link the optimal functioning of your brain to the activity of your muscles begins just after the Earth formed, about 4.6 billion years ago. The first life forms appeared rather quickly, as soon as the Earth cooled to the point that water remained in a liquid state. That is why scientists at NASA are always excited to find planets or moons with liquid water on them. Water, rocks, and heat are all that are necessary for life to arise. The first "cells" that formed on the young Earth were very simple; they survived and thrived because they could produce their own energy and replicate. These primitive cells were similar, but not yet identical, to what we know today as bacteria. The descendants of these simple single-celled organisms are still the dominant life forms on our planet. The top five miles of the Earth's crust, and possibly much deeper, is thick with them; they also thrive throughout the entire atmosphere.

About 3.2 billion years ago, thanks to the process of photosynthesis by cyanobacteria and blue-green algae, the oceans became saturated with oxygen; different forms of life took advantage of this highly reactive, and also very toxic, gas in order to enhance their ability to multiply and survive. These primitive cells had the same goals as your brain does today: to survive in order to multiply. The winner of this battle of life was, and forever will be, determined by whoever makes the most copies of themselves for the longest period of time. Whatever life form failed in these goals quickly became extinct.

Then, about 2 billion years ago, some of these simple cells began to organize themselves into complex multicellular groups that evolved the ability to move around their environment. Although our species appeared on the scene only about 200,000 years ago, we share the same goal as all of the other species that preceded us: to survive in order to procreate. Both of these activities require movement.

### Oxygen: the devil's Janus coin

Movement requires energy. The first simple cells that appeared were not very adept at making energy. They produced just enough for their own survival and replication; not much was left over for other purposes. One of these single-celled free-living organisms evolved a brilliant biochemical trick that allowed it to utilize the abundant oxygen in the air for the production of lots and lots of energy. One day, it was a Saturday in late April, about 1.8 billion years ago, this oxygen-eating simple cell, today called a *mitochondrion*, was engulfed by a somewhat larger single-celled organism and the two developed a wonderful symbiotic relationship where the smaller cell, now safely nestled inside the bigger cell, provided lots of energy that enhanced their joint survival. This additional energy could now also be used for one additional ability—*movement*.

The first appearance of simple muscle-like proteins that could change shape by utilizing this additional energy source probably occurred about 640 million years ago in simple sponges. These first muscles were involved in cell motility and survival by regulating the circulation of seawater to harvest nutrients. Muscles for movement have always been critical for obtaining food for survival. Thus, the biochemical links between muscle function and energy utilization evolved concurrently. The first muscle protein was a type of myosin. Myosin is a critical component of muscle fibers found throughout nature and in the muscles of humans. The large muscles of your body were the result of a simple process of duplication and modification of the same genes that initially evolved in these ancient sponges.

Survival, and the successful procreation, of bigger and more complex organisms was greatly enhanced by the ability to move toward sources of food and away from predators. Thanks to the mitochondria, whose ancestors gave up their independent lifestyle, and their remarkable ability to utilize oxygen, considerable amounts of additional energy were now available for the development and activation of primitive muscle cells that could move the entire organism around its environment. Self-powered movement was a significant step forward in the evolution of more complex organisms.

Once again, the devil was in the details. The abundant supply of energy came with one important condition: Every cell now required a constant, unending, and invariable supply of oxygen. Fortunately, oxygen was abundant in the air. Unfortunately, oxygen is incredibly toxic to cells. I know that sounds crazy; everyone was taught in school that oxygen is critical for life. It is, and thanks to hemoglobin in your red blood cells, the oxygen is transported safely to all of your tissues. The role of hemoglobin in your blood is to bind oxygen tightly to itself in order to make a sufficient number of oxygen molecules available for the process of respiration inside each of the thousands of mitochondria inside every one of the 37 trillion (approximately) cells of your body. (The 900 trillion bacteria, molds, and fungi that coexist inside your body also utilize the oxygen that you inhale.) Hemoglobin's other job is to keep the concentration of oxygen low enough that it does not kill every one of your cells in the process.

This devil of a deal, with oxygen as the currency, resulted in at least three fascinating consequences. First, it is the reason that there are only two sexes on this planet (I am ignoring the kingdom of fungi and a recently discovered worm—they discovered some interesting alternative solutions). Second, it explains why males of all species never live as long as females of all species (more on this later). Third, it explains why all complex cells, including you, die. Prior to this devilish deal between two simple cells billions of years ago, sex and death did not exist. For at least 2 billion years after life first appeared, cells just grew bigger and bigger until they divided into two "daughter" cells. Unless an errant space rock fell on it, these primitive cells never died. There was also no sexual reproduction; these first simple cells were neither male nor female. Now, billions of years later, big, complex, multicellular, muscular humans age and ultimately die because of this ancient devilish deal with oxygen. Today, we all require oxygen to metabolize our food in order to survive, procreate, and exercise. Right now, while you are thinking about how to achieve these important goals, your brain is using a lot of oxygen. I know that it's risky, but please do not stop thinking.

Active brains require a lot of energy. As simple nervous systems evolved and became more complex, bodies evolved longer and longer intestines in order to optimize the extraction of more energy from the diet. It is not surprising to discover that, for mammals at least, the length of the gut is significantly correlated with the total body mass. Initially, this evolutionary modification was able to satisfy the increasing demands of bigger bodies. Brains use a lot of energy; thus, you might predict that big brains lead to the evolution of longer intestines. However, a study of over 100 different mammals did not find any correlation between the size of the brain and the length of the gut. Unexpectedly, there is a negative correlation between brain size and total body fat for most mammals—but not for humans. We humans have big brains, bigger than other primates, and we are fatter too. Why? Maybe our big human brains require lots of body fat in order to function optimally. At first glance, this is a reasonable expectation given that the brain requires a constant and uninterrupted supply of energy. Unfortunately, this explanation is not correct because, under normal circumstances (unless you have not eaten for a couple weeks!), the brain does not obtain its energy from fat. Whatever pseudoscience explanation you might have read about related to the ketogenic diet is total nonsense. Optimal brain function requires sugar above all else. Your feeding centers are *only* capable of detecting sugar in your blood, not fats or proteins. That's why it is not very satisfying to eat a meal that does not contain any simple carbohydrates. Never forget, your brain will often prefer that you eat things that your body would prefer you did not eat: blame that on evolution. Many more examples of this brain feature will appear in the following chapters.

Overall, humans have a big brain, a fatty body, and a gastrointestinal system that is fairly efficient at extracting energy for itself and its two principal customers, the reproductive system and the brain. As you can easily appreciate, muscles were never a big part of the evolutionary equation. However, humans have lots of muscles; they are critical for our survival and procreation and they need to be fed.

The quality of the diet, and the increased availability of nutrients, of our ancestors was also greatly enhanced by their newly discovered ability to control fire and cook their food. This advance appears to have occurred sometime between 500,000 and 1.7 million years ago. Exposing food to heat before consuming it allows our intestines to extract more nutrients from our diet, especially meat. Meat is an energy-rich source of nutrients and minerals. Around this same time a few significant genetic changes occurred in the expression of some gut enzymes that allowed our ancestors, and now us, to extract more energy from our cooked food in order to support movement, and ultimately exercise. In the next chapter I will examine how we obtain energy from our food in order to support exercise because the process of obtaining energy has a consequence on brain function.

# 4

# MAKING ENERGY FOR EXERCISE

In terms of weight, skeletal muscle is typically the most abundant organ of the human body. Eighty percent of the weight of each muscle is water. Although your muscles are mostly water, they consume a lot of oxygen, overall about as much as your liver or brain. Their basic function encompasses the maintenance of postural support, the generation of force and power during voluntary movements, and breathing. Muscles also play a major role in keeping us warm; this is called *thermoregulation*. Skeletal muscle is also responsible for a large portion of the oxidative metabolism. Oxidative metabolism is a chemical process by which the oxygen you inhale is used to make energy from the food you eat; heat is produced as a byproduct of this metabolism. You might have noticed that a few hours after eating your body feels warmer. People who consume a calorie-restricted diet tend to feel cold most of the time. Your body heat is mostly generated by the chemical reactions taking place within your liver that is processing your food for absorption into all of your human cells as well as the many trillions of nonhuman cells that share your body. Because free oxygen from the air is utilized in this process, it is called *aerobic metabolism*. Aerobic respiration is a set of metabolic reactions and processes that take place in your cells to convert the energy stored inside food into adenosine triphosphate (ATP), and then release waste products. What makes making ATP so important that every cell in your body spends all of its time and energy producing it?

### ATP

You should think of ATP molecules as the tiny batteries your body requires in order to stay alive. You must constantly make them because you are constantly using them up. They are critical. When you stop making them—you die. When someone takes cyanide in order to commit suicide, the reason they die is because the cyanide molecules prevent their cells, all of their cells, from making ATP. Death usually takes about 30 minutes (depending on dose and body size) and, due to the cellular mechanisms underlying the death of each cell, the process is extremely painful. For some reason, whenever spies die in an action movie by hurriedly crushing a cyanide pill in their mouth, it always appears to be fast and painless. It's neither.

Muscles are also a major storage site for energy-rich nutrients, such as glucose, simple fats, and a large variety of amino acids; thus, your muscles play an essential role in coordinating whole-body energy production and metabolism. For example, amino acid metabolism in muscle rapidly adapts in response to physical exercise, the type of dietary protein available to the muscle, or the presence of hormones such as insulin-like growth factor-1 (IGF-1) or testosterone. IGF-1 is primarily produced by the liver. IGF-1, as its name suggests, is very similar to insulin and its release from the liver is induced by milk as well as other dairy products in the diet. This reaction to dairy may be related to our need, since humans are mammals, for milk during the first few years of life. Why humans continue to drink the milk of other (!) mammals into adulthood is likely related to culture rather than a biological need. No other mammal drinks the milk of another mammal—only humans do this (and the occasional Jedi Master living in exile on the aquatic planet of Ahch-To).

Getting back to IGF-1. IGF-1 can cross the blood–brain barrier and might act as a mediator of the exercise-induced changes in the brain (I will return to this important ability later). IGF-1 controls the release of another important hormone called *growth hormone*. Together, testosterone and growth hormone control muscle development and growth, particularly during childhood. It was once thought that muscle growth after weightlifting exercise was due to the combined actions of growth hormone and testosterone; however, recent studies have raised significant doubts about their combined impact on muscle growth after lifting weights.

### A brief PSA

Please allow me to state the obvious at this point, given our general ignorance about the actions, and interactions, of your body's hormones: It is never wise to take supplements of these hormones. Inducing an imbalance—either too much or too little of them—can produce devastating long-term harm to your body. This advice also applies to any attempts to compensate for age-related losses of specific hormones. Don't do it! The balance of hormones and their interactions with each other are simply too complex to risk the consequences of consuming megadoses of any particular hormone supplement. Purveyors of supplements on the internet have no interest in your good health; they are far more interested in increasing the size of their purse than the size of your muscles.

### The importance of ATP

Your body consists of around 600 muscles that may contribute about 40% of your total body weight. As our understanding of muscle physiology advances it is becoming clear that muscles behave as though they are a complex endocrine system, much like your thyroid gland or pancreas. Muscles send out chemical signals so that the brain and body are aware of what they are doing and their current metabolic needs. However, unlike all of your other endocrine organs, your muscles can contract and dramatically change shape. Surprisingly, although the contraction process of muscles has been studied for many years, the molecular details of the contraction process are still not fully understood. What is well known is that muscles require the energy stored in molecules of ATP for contraction.

ATP plays a critical role in every part of your body and is necessary for you to exercise. It is an ancient molecule that existed naturally on the planet long before any life appeared. The first primitive cells simply exploited whatever suitable molecules were lying around and incorporated them into their own chemistry. The constant demand for so many molecules of ATP, and your body's production of them, primarily via the metabolic activity inside your mitochondria, has consequences on your brain function, your general health, and how well you age. Especially if you are a male (more on this later).

### Mitochondria make ATP

Mitochondria are small, membrane-bound organelles that, as I described earlier, were once free-living simple cells that now reside inside virtually every cell of your body. The mitochondria are essential for the production of ATP that your cells use for many critical chemical processes, including muscle contraction; however, the constant moment-by-moment demand for ATP has consequences on the quality of life and ultimate death of humans.

Beginning about 3 billion years ago, the forces of evolution linked three vital processes together—movement, energy production, and death—and the nature of life on this planet changed forever. The biochemical power plant that achieves this linkage is the mitochondria. Your muscles contain thousands of mitochondria. In fact, muscles contain more mitochondria per weight than any of your other tissues. Mitochondria have their own DNA and multiply independently of the cell in which they reside. They behave as though they still believe that they are the free-living, independent organisms that they once were so very long ago. Over time, our cells modified how the mitochondria function so that their primary function now is the generation of large quantities of energy in the form of ATP. They are the power cells of your cells.

Mitochondria use the carbohydrates, fats, and proteins from your diet to generate ATP. Glucose is a carbohydrate and is the simplest form of sugar. The adult brain has a very high energy demand requiring the continuous delivery of glucose from blood. The brain accounts for approximately 2% of the body weight but consumes approximately 20% of glucose-derived energy, making it your body's foremost consumer of glucose. The largest proportion of energy in the brain is utilized for neuronal computation and information processing—that is, thinking.

Obviously, your brain requires a lot of sugar; without a constant uninterrupted supply you will quickly lose the ability to think and slip into a coma. You obtain most of this sugar from your diet. Sadly, somewhere in our evolutionary history, we lost the ability to convert fat into sugar; unlike a few lucky creatures, humans cannot perform this metabolic trick. Humans can make glucose from muscle protein, but we need to break down the muscles, and this process requires lots of ATP. From your brain's perspective, dietary sugar is indispensable. It will do whatever is necessary to convince you to eat sugar as often as possible. I like to remind my students that sometimes your brain will demand that you consume certain foods and drugs that are not always good for the rest of your body. Sugar is an excellent example of this irony. The brain loves sugar and will reward you with feelings of euphoria for eating lots and lots of sugar. Unfortunately, eating lots of sugar every day will ultimately lead to obesity and diabetes and predispose you to developing cancer. Needless to say, your brain also really loves fat and salt. As everyone is well aware, the brain always wins this contest for control of what gets eaten. Why? Because fat, salt, and sugar occurred very rarely in the environment during the evolution of brains. Bodies require fat, salt, and sugar; thus, our brains evolved neurological mechanisms that reward us whenever we consume them. Thus, today, we are all stuck with a fat-, salt-, and sugar-loving brain. Unfortunately, the blind processes of evolution cannot work backward to correct what has become a major risk factor for long-term health.

So, in summary, you eat and breathe and your mitochondria provide energy for your muscles by metabolizing the carbohydrates, fats, and proteins that you consume. The chemical machinery inside the mitochondria then gobbles as much energy from the process as possible for the production of ATP. The fact that your mitochondria accomplish this task so inefficiently means that much of the energy in your food is lost as heat. The heat generated by the thousands of mitochondria within muscles mostly explains why heavily muscled male bodybuilders, with their tissues rich in testosterone, like to wear so little clothing on cold winter days when the rest of us are shivering. It has absolutely nothing to do with their egos!

# 5

# EXERCISE REQUIRES EATING AND BREATHING

Your mitochondria are miniature carbon-burning factories. The fats, carbohydrates, and proteins that you eat every day are mostly made of carbon atoms linked together by what chemists call *bonds*. Your mitochondria contain enzymes that break apart these carbon bonds; in the process, the energy is released and used to produce the ATP you need to survive and your muscles require in order to contract. Once the energy is captured, your mitochondria have a waste disposal problem: What are they going to do with all of the leftover carbon atoms? Billions of years ago, they found the answer! Combine the leftover carbon atoms with the readily available gas—oxygen—and expel the product as a gas called *carbon dioxide*. Oxygen in, carbon dioxide out, is the essence of breathing. Voila! Problem solved.

Not quite. Unfortunately, this easy solution for discarding carbon introduced another, far more serious problem. Oxygen is toxic; it must be handled very cautiously. In general, the hemoglobin in your blood does a decent job of regulating the oxygen levels near the individual cells of your body so that your cells have the oxygen they need for removal of the carbon debris, but not so much that they risk being killed outright. In response to all of this oxygen permeating your tissues, your body evolved numerous "antioxidant" systems that allow you defend yourself against the oxygen. That is why they are called *antioxidants*; they prevent injury from oxygen. Having lots of antioxidant systems allows us to live a long healthy life. How long? It is currently thought that these antioxidant systems allow our species to have a maximum life span of about 117 years. Obviously, most of us do not live that long. Why? Part of the answer is that we eat all of the time, and move around all of the time, and therefore must keep breathing to remove the carbon debris; this makes us vulnerable to the consequences of oxygen permeating our body. If you could only stop eating and breathing . . . well, you would die. There is no way out of this conundrum. Remember what I said earlier: Because of your single-celled ancestor's devilish deal with the mitochondria billions of years ago, you will die.

### Reactive oxygen species

While you are eating and breathing in order to survive long enough to reproduce the next generation of eaters and breathers (my future students), tissue-damaging molecules called *reactive oxygen species* (ROS) are constantly being produced by the mitochondria that live inside virtually every cell of your body. The story is not all bad; ROS are not always harmful. Oxygen has been around a very long time; during that time your body, as well as your pet's body (my students always ask me if these rules also apply to their cats, dogs, birds, etc.), evolved some effective chemical tools to defend you from ROS—unless, of course, you eat too much and move too much and therefore produce too many ROS. Unfortunately, too many ROS can overwhelm your natural antioxidant systems, slowly destroying your neurons as well as every other cell in your body. That is why you are always being encouraged to eat foods that offer additional antioxidant benefits, such as colorful fruits and vegetables. If you would like to see a visual representation of the actions of ROS please watch my TED talk here: https://www.youtube.com/watch?v=4SvkaK2Aloo&feature=plcp.

Think about the unbelievable irony of this process. The mitochondrial power plants that reside in quite large numbers in every cell of your body, particularly your muscles, are actively injuring those same cells that they reside within by the very process that keeps them alive. A mitochondrion is the original Trojan horse and the ROS are Odysseus and his men hiding inside just waiting to emerge and do lots of damage to Troy (your body). It turns out that each species' maximum lifespan may be determined by how many ROS are produced by the hundreds of mitochondria that live in each of their cells. Thus, even at the level of individual cells, Cicero was correct: We are our own worst enemy. Overall, for all tissues in your human body, for all tissues of all animals on this planet (except for a rare genus of *Monocercomonoides* that do not contain mitochondria), mitochondria are a principal source of harmful ROS that underlies aging. There is simply no way around this; we live on a planet bathed in oxygen.

If you think this is bad news, the situation is much worse for males. The biochemical processes that occur within the mitochondria are responsible for aging for both males and females; however, the mitochondria in females produce significantly less ROS than those from males. The damage to mitochondrial DNA due to the elevated levels of ROS is four times higher in males than in females. Thus, the mitochondria in males age their hosts faster than those within females. It should eminently be clear by now that we require lots of healthy mitochondria in order to stay alive! Women are protected from many of the negative actions of ROS by their higher levels of estrogen as compared to males (at least until menopause, when men tend to have higher levels of estrogen than women).

It's all about the ovaries and their continued good health. Ovariectomy abolishes the gender differences between males and females and estrogen replacement rescues, under most circumstances, from the ovariectomy effect. The acceleration of age-associated processes and the increased vulnerability of females to a large variety of degenerative diseases that accompany the onset of ovarian failure at menopause are due to falling estrogen levels and increased vulnerability to elevated levels of ROS and pro-inflammatory proteins in the body and brain. Menopause allows the indispensable act of respiration to reproduce the same pro-inflammatory, and quite harmful, environment in the brain of older women that is present in older males. Due to these menopause-associated changes in body chemistry, women enter a phase of their life when they are at increased risk for disorders associated with extensive brain inflammation, such as depression, bipolar illness, and schizophrenia, as well as becoming more vulnerable to developing age-related disorders such as Alzheimer's disease, Parkinson's disease, and amyotrophic lateral sclerosis.

This symbiotic relationship with mitochondria that offers so many advantages also came with a huge price tag—the ultimate death of any cell that possesses mitochondria. That last statement requires some explanation. The single most important factor that increases your chances of dying is causally related to how many days you have been alive. I know that statement sounds totally obvious, but it does raise an important question: What is it that you do every day of your life that increases your chance of dying? The answer is that you eat and breathe—every day. Eating provides your body and brain with the energy stored within the carbon bonds that are contained within the fats, carbohydrates, and proteins that make up your diet. Breathing brings oxygen to your mitochondria to carry away the carbon debris that forms when these bonds are broken apart. This single critical activity, called *oxidative metabolism* or *respiration*, that is essential for your daily survival, is the most important factor that very slowly, minute by minute and day by day, ages you until you die. That's why consuming fewer calories every day is the only effective way to slow the aging process. Eating antioxidant-rich fruits and vegetables is also helpful.

## The double-edged sword

The preceding paragraphs about the vital role played by mitochondria explain why any consideration about the benefits of
exercise on your brain and body must consider the role of oxygen: your critical need for it while exercising balanced against
its toxic effects in your body that cause you to age just a little
bit every day. When you exercise moderately every day, your
muscles, and the millions of mitochondria that dwell within
them, will adapt to the increased intake of oxygen. Because
of mitochondrial adaptation, exercising presents itself as a
double-edged sword. Moderate exercise adapts muscles to the
consequences of the increased metabolism required to support it. The body has evolved the ability to adapt to increasing
physical activity by increasing muscle size and strength while
simultaneously improving our ability to obtain and deliver
more oxygen to our tissues. In contrast, too much exercise, particularly for a long period of time, pushes metabolic demands
to the point that your cells' innate protective systems are overwhelmed and the increasing levels of ROS lead to accelerated
tissue injury and aging. For the brain, as compared to the other
tissues of your body, the abrupt increase in ROS levels associated with extreme levels of exercising is much more harmful.
The brain is exceptionally vulnerable to increased ROS formation due to its own very high metabolic rate. Our fast, smart
brains came with their own evolutionary price tag.

## KGDHC

Fortunately, daily moderate exercise reverses the negative consequences of oxygen by reducing the production of one particularly nasty ROS, hydrogen peroxide. This is how. Mitochondria
possess protective enzymes. One of these has the alphabeti-
cally challenged name KGDHC. KGDHC is readily induced
by exercise and inhibited by hydrogen peroxide. When it is
not functioning adequately, ROS production and hydrogen

peroxide levels increase. This complex relationship between exercising, the production and utilization of energy, the mitochondrial requirement for a constant supply of oxygen, and the subsequent exercise-induced changes in muscle chemistry has only recently been investigated. Thus, the story is still evolving as more knowledge is gained about these processes.

Scientists have discovered that the genes that control energy metabolism in muscles have been highly conserved across millions of years of evolution. Essentially, the better you negotiate your energy–oxygen interchange with your indwelling mitochondria while they provide you with the ATP that is required for movement, the healthier you are both physically and mentally. There are significant negative consequences in meeting the demands of continually exercising muscles. Fortunately, the activities of many different oxidative damage-repairing enzymes, such as KGDHC, are increased with exercise training. But not always—it all depends on the nature of the exercising.

### The consequences of making ROS

Is exercising good or bad for you? To find an answer, we need to return to the principle of moderation, a concept that I have been emphasizing since almost the first page. This concept has been around at least since the time of the ancient Greeks, who wrote the inscription *Meden Agan* (μηδὲν ἄγαν) over the Temple of Apollo at Delphi. It warned "Nothing in excess." Given that their culture emphasized sport, I would like to think that this warning also applied to their athletes. Then, as now, it is necessary to find the right balance. We should all discover our personal balance between getting just enough exercise to produce benefit while avoiding the harmful consequences of eating and breathing that are necessary in order to exercise.

Here's some surprising, and at first glance rather contradictory, good news: Regular moderate exercise increases the production of ROS. Fortunately, this leads to some critical adaptive and beneficial responses by your muscles. This adaptive response happens because moderate levels of exercise induce the birth of new mitochondria. Having more mitochondria leads to greater levels of protective enzymes such as KGDHC. In contrast, extreme levels of exercise are not associated with the birth of new mitochondria or the presence of additional protective enzymes, thus exposing the brain and body to prolonged and toxic levels of ROS. You might expect that having more mitochondria would lead to the production of greater levels of toxic ROS. In fact, having additional mitochondria to assist with muscle contractions during regular exercising means that each individual mitochondrion is allowed to work at a lower respiration rate. Essentially, there are more of them to share the workload. Thus, each of them generates much lower levels of toxic ROS molecules. Overall, thanks to all of your newborn mitochondria, as long as you maintain a steady regimen of moderate exercising, you do not produce as much ROS and the natural aging process is not accelerated significantly. Notice what this conclusion does not predict: Exercising is *not* going to slow down normal aging. Modest exercise, such as a daily stroll in the park, will simply allow you to age more slowly than extreme exercising, such as running 10 miles every day. Unless you happen to be a male, because . . .

# 6

# IT'S NOT GOOD TO BE A MALE

Many of my male students participate in some form of daily exercise routine. Every year I warn them about the consequences of too much exercising, but they are young and feel immortal and usually choose to ignore my advice. Being a male, or more accurately having lots of testosterone circulating through the body, whether you have lots of big muscles or not, complicates the impact of exercise and energy production on mitochondrial function and your body's health. Testosterone increases the energy expenditure from muscle mitochondria, leading to an increase in the number of mitochondria within each muscle and an increase in the production of ATP. Much of the energy used for the increased production of ATP is drawn from the body's stored fat.

So far, this all sounds OK. But the hormone-driven mechanisms that control the metabolism of fat for energy have significant consequences for males. Testosterone alters how males metabolize food and increases the amount of heat their muscles produce during normal respiration. Testosterone, due to its effects on a specialized protein called *uncoupling protein* or *thermogenin*, makes the normal food-to-energy conversion process in mitochondria become inefficient—that is, cells waste more energy as heat, making men feel warm. Lacking both testosterone and significant muscle mass (typically), women tend to produce less body heat from their food; consequently, it is usually much harder for women to lose weight than it is for men. The male body, particularly all of those muscles, is capable of wasting a considerable number of consumed calories as body heat. This explains why the principal method of dieting for many of my male students is to eat anything they want and then exercise off the calories. Or, they could do nothing at all: Males, to the chagrin of many females, can lose weight by simply sitting still. In contrast, women, who lack the mitochondrial uncoupling actions of testosterone, are forced to lose weight the old-fashioned way, by not consuming as many calories.

For males, wasting calories in order to produce heat has some negative long-term consequences. First of all, males need to consume more calories per day than do females; consequently, males generate more harmful reactive oxygen-free radicals—ROS. These ROS are quite harmful to the body and negatively affect men's health and reduce their longevity compared to women.

It's not just human males who have reduced lifespans due to eating and breathing. Human males do not necessarily lead more reckless lives than do females, at least not after age 30. Males of *all* species, including spiders, flies, and the birds and the bees, never live as long as females of all species. This is true even for species with a very short lifespan, such as a fly. Male flies of one species have an average maximum lifespan of five days; the females of the same species have an average maximum lifespan of seven days. In contrast, women waste less energy as heat, need to consume fewer calories every day in order to survive, and produce fewer ROS, all of which benefits their overall brain health and longevity.

If ROS are so harmful, why does your body produce them? Because you have no choice. If you have stopped making ROS, you are probably dead.

# Part II

# YOUR BRAIN ON EXERCISE

# 7

# BDNF

## MUCH ADO ABOUT SOMETHING?

Assuming that your brain does care whether or not you are exercising, how does it know? Actively contracting skeletal muscles communicate with the brain, indirectly and directly, via chemical messengers in the blood. One of these potential chemical messengers that has attracted lots of attention is called *brain-derived neurotrophic factor*, or *BDNF*. Is BDNF the long-sought-after chemical messenger that communicates between active muscles and the brain? Based on limited evidence, a large number of popular articles and books have claimed that the beneficial actions of BDNF on the brain are so amazing that it alone justifies exercising. These articles and books claim that BDNF will make you smarter and cure your depression. Such outrageous claims, especially when based on highly preliminary findings, are almost always proven incorrect sooner or later. The reality of BDNF is much more complicated. When it comes to understanding the role of BDNF in the benefits resulting from exercise, the devil is most definitely hidden deep within the details.

BDNF controls bodily functions that are indirectly related to exercising, such as regulating energy homeostasis by controlling patterns of feeding and by modulating glucose metabolism in peripheral tissues. Via these actions, BDNF may mediate the beneficial effects of the energetic challenges of vigorous exercise by stimulating glucose transport into muscle cells and their numerous mitochondria. These are just some of the beneficial effects that BDNF provides to the body. However, BDNF can also induce harmful actions in your body, depending on your current physiological state. For example, BDNF plays a pivotal, and decidedly unfortunate, role in the growth, survival, and chemoresistance of tumor cells in various types of cancers. If you consider the substantial nutritional requirements of fast-growing cancer cells, it makes sense that BDNF would play a role in providing energy for their growth. BDNF is simply performing the job that it evolved to do for the rest of your body.

Most muscles produce BDNF during aerobic exercise. The amount of BDNF produced varies depending on the type of muscle, the specific activities, and the intensity of exercise training. For example, low- to intermediate-intensity movements usually activate more slow-twitch than fast-twitch muscle fibers. For example, treadmill training increased BDNF levels in the soleus muscle (containing slow-twitch fibers), but no changes in the gastrocnemius muscle (containing fast-twitch fibers) were observed.

BDNF levels also increase significantly in the blood following repeated aerobic exercise. The increase in BDNF blood levels requires repeated exercising since no significant effects on BDNF levels were detected after only one bout of exercise. These early findings were consistent with the idea, since proven wrong, that exercising muscles are the source of BDNF in the blood. Muscles do produce BDNF during repeated exercising, but they do not release their BDNF into the blood. The BDNF produced by muscles is utilized to stimulate the conversion of fat into energy by the mitochondria within muscles. Thus, muscles use their own BDNF to perform processes related to their own energy utilization.

What, then, is the source of the BDNF in the blood following prolonged exercising? The BDNF in blood after exercising is most likely released from activated platelets. Platelets hold about 90% of the BDNF found within whole blood. The essential trigger that induces the release of BDNF from platelets into the blood following exercising is the cytokine interleukin-6 (more on this molecule later). Thus, one possible hypothetical scenario goes as follows: Repeated exercise elevates blood levels of interleukin-6, which induces platelets to release their BDNF; the BDNF is then available to enter the brain. This would be a workable explanation, if it were all true—but the BDNF found in blood cannot enter into the brain because of the presence of the blood–brain barrier.

For example, when BDNF was injected into the blood, it did not have any beneficial effects, such as neuroprotection from ischemia or injury, inside the brain as long as the blood–brain barrier was intact. That last point is important because some authors have claimed that BDNF does enter the brain. Unfortunately, the animals used in these older studies most likely did not have an intact blood–brain barrier. In animal studies, BDNF is only neuroprotective if it is injected directly into the brain, thus bypassing the blood–brain barrier. One recent study did discover a way for BDNF to get across this barrier: The BDNF was chemically attached to another molecule that acts like a Trojan horse to fool a transporter system into moving the BDNF across the blood–brain barrier. Unfortunately, attaching a Trojan horse molecule to the BDNF in your blood is not an option for human athletes.

Thus, whatever the biological mechanisms that ultimately induce platelets to release their stored BDNF following repeated exercising, the goal is not to influence brain function directly. Therefore, the exercise-induced change in blood levels of BDNF must be investigated for potential consequences within the body that hopefully have downstream, indirect beneficial consequences within the brain.

### Addressing the BDNF myth

You do not need to search too far to find claims that exercise will make you smarter because it increases BDNF levels in the brain. Does it? There have been numerous human and animal studies exploring the relationship between exercise, BDNF, and cognitive performance, and these studies have produced very mixed results. The confusion is partly due to methodological differences in the way each investigator designed the exercise routine; the particular cognitive ability being determined; and how, where, and when BDNF levels were measured in relation to the exercise routine.

I want to begin with how the BDNF was measured in the blood. First of all, most scientists collect samples of blood from their subjects and then extract either serum or plasma for analysis. For BDNF, this is where the confusion begins. Serum and plasma differ in many important ways; thus, conclusions drawn from studies using serum differ from those studies using plasma. I know that sounds rather odd; shouldn't both contain the same amount of BDNF? No, they do not. Serum and plasma are both derived from the liquid portion of your blood that remains after the cells have been removed. Serum is the liquid that remains after the blood has clotted. Plasma is the liquid that remains when clotting is prevented by the addition of an anticoagulant. I know those differences sound trivial; however, the devil is again hiding in the details, and it really matters in this case that you understand why.

Serum contains much higher concentrations of BDNF than plasma; the concentrations of BDNF in serum and plasma may differ by a factor of 200! In order to obtain serum, blood samples must be allowed to coagulate prior to preparation. During the coagulation process the platelets in blood spontaneously release their storehouse of BDNF into serum. Please recall that almost all of the BDNF found in blood is stored in platelets. It turns out that the longer the technician allowed the blood to coagulate before preparation, the greater the levels of BDNF were detected. If you look at the clotting times in the published literature, they varied greatly between different BDNF assays. Obviously, this coagulation effect does not happen with the preparation of plasma.

This difference in preparation explains why there is no correlation between serum and plasma BDNF levels. Despite the obvious differences and high variability of results between serum and plasma BDNF, these measures are used interchangeably in human and animal literature, and this has led, in my opinion, to many of the inaccurate conclusions about the hypothesized linkage between exercise and BDNF. For example, the problems associated with measuring BDNF underlie the confusing set of findings that claim a positive correlation between your level of physical fitness and the amount of BDNF in your serum that actually flips to become a negative correlation if the time allowed for your blood to clot goes from 30 minutes to 60 minutes, respectively. Thus, if you would like to spin a story that claims a correlation between exercise and BDNF in the blood, the data are available to do so. However, if you would prefer to spin the opposite story—that there is no correlation—the data are available to do so.

It gets worse. As I continued to investigate the relationship between BDNF and exercise, I discovered that there are some additional confounding factors that most investigators rarely took into consideration. The level of BDNF in the serum is influenced by race (blacks have more) and sex (females have more). The number of platelets per volume of blood can also influence how much BDNF is measured; more platelets mean more BDNF will be found in each sample. Unfortunately, platelet number can range from 150 million to 450 million cells per milliliter. In addition, platelet number is easily altered by factors rarely noted in most studies, such as tobacco use, depression status, or the consumption of certain common foods such as alcohol, dairy products, or cranberry juice.

The BDNF protein in plasma samples is also unstable: It tends to degrade and fall apart into its individual parts while being stored. Furthermore, it depends on when the blood is sampled following exercising. The problem is that the BDNF protein, once released into the blood, can only survive for less than an hour before being destroyed by enzymes. Thus, if you wait too long after exercising to take a blood sample, the levels of BDNF will be artificially low.

Proving that a correlation exists between BDNF and exercising is even more difficult because plasma BDNF levels tend to be lower in older or heavier patients with elevated levels of low-density lipoprotein (LDL) cholesterol. Older and overweight subjects, with their lower levels of BDNF, tend to show up more often in the control group of population studies because they do not exercise as much. This will bias the outcome of the study to support a false positive relationship between BDNF and exercising. Furthermore, the age- and weight-related declines in BDNF plasma levels were seen even though the number of platelets did not change. Thus, a person's age may induce blood platelets to produce significantly less BDNF, leading to reduced blood levels that are independent of whether or not the person exercised.

It gets worse. Complicating these investigations even further is the controversy over whether plasma BDNF levels demonstrate retest stability when measured twice within one year in the same person, combined with the substantial circadian changes in background levels of BDNF throughout the day. This makes replicating earlier results in different laboratories very difficult. Taken together, it is likely that the initial excitement about BDNF was due to the ease of assaying BDNF in the blood rather than it being a good candidate for the communication between exercising muscles and the brain.

### BDNF in brain and blood

Despite the numerous methodological concerns raised in the previous paragraphs, too often some authors have claimed that changes in circulating BDNF levels may be used as an indicator of changes occurring in the brain. The available scientific literature does not offer an easy resolution to this issue. A few studies have reported a positive correlation between the blood and brain BDNF concentrations, but others have found no relationship. Once again, some of the confusion is related to whether the investigators measured plasma or serum levels of BDNF. The best current evidence suggests that the level of BDNF in the blood does not correlate with the level of BDNF in the cerebrospinal fluid. The concentrations of BDNF are much higher in the serum (more than 1,000 times) than in the cerebrospinal fluid. You might expect that this incredible concentration gradient would literally push the BDNF through the blood–brain barrier, but, as I discussed earlier, BDNF is simply too large and too electrically charged to get across the blood–brain barrier.

Complicating the correlation between exercising and blood levels of BDNF is that fact that the BDNF has so many sources in the body that probably have little to do with exercise or the brain. For example, your immune system, liver, and vascular endothelial cells are all potential sources of BDNF in the blood. BDNF measured in blood may reflect changes in function within these tissues rather than whatever is changing BDNF levels due to exercise. Overall, available evidence indicates that it is unlikely that peripheral changes in BDNF levels will translate as a useful biomarker of brain BDNF levels.

### Blood BDNF and depression

Some early reports have claimed that a lack of physical activity lowers blood levels of BDNF and induces depression. Thus far, we have only correlational findings that link serum BDNF with depression. For example, BDNF levels in serum are significantly decreased in depressed patients, and this decrease is normalized by antidepressant treatments such as one of the popular serotonin or norepinephrine reuptake inhibitors, electroconvulsive shock therapy, and daily medical marijuana. Thus, if exercise does alleviate depression, these publications claimed that it may act via similar mechanisms to these antidepressants. Once again, these initial reports led to meretricious claims about the curative effects of exercise on depression. While exercise is beneficial, it should never replace, or delay, the use of standard pharmacological treatments.

In contrast to these past reports, more recent investigations have *not* found a correlation between serum BDNF levels and depression, which would be expected if serum BDNF played a direct role in controlling depressive symptoms. The problem, once again, is that serum BDNF levels are dependent on the release of BDNF from platelets, which was not evaluated in studies of depressed patients. Another aspect to consider is that other diseases, such as schizophrenia, bipolar disorder, and anorexia, among others, are associated with decreased levels of serum BDNF. Thus, reduced plasma BDNF is not a specific diagnostic marker for depression. However, this does not rule out the possible role of BDNF in these neurological conditions.

Surprisingly, some studies have reported that plasma BDNF levels were significantly higher in the depressed patients. Thus, lower plasma BDNF concentrations seen in depressed patients and the increase in BDNF levels observed during the course of antidepressant treatment may be a secondary effect that does not causally explain the recovery from depression. The correlation between peripheral BDNF levels and depression is probably related to the tendency of platelets to release their BDNF in response to some unknown peripheral signal. Therefore, an important question that remains to be answered is whether the relationships found between exercising, serum BDNF levels, and depression might somehow be mediated by other chemical signals from muscles. The next chapter investigates the potential roles of some of these chemical messengers.

# 8

# MUSCLE SIGNALS
# TO YOUR BRAIN

There are many molecules other than BDNF that originate in the periphery that might directly or indirectly mediate the beneficial effects of physical exercise on the brain. The complex blend of chemicals released from exercising muscles enhances a broad range of brain functions, including improved vascularization, neuroplasticity, memory, sleep, and mood. Let's take a look at some of the most interesting of these.

## Lactate

One of the best-studied molecules produced by muscles during normal exercise is lactate. Under normal resting conditions, when you are not exercising, the amount of lactate produced in your body is usually equal to the amount metabolized by your tissues for energy production. In response to exercise, because it is one of the chemical byproducts of glucose metabolism, lactate is found at higher levels in the blood. Although lactate cannot cross the blood–brain barrier easily on its own, it is capable of being transferred into the brain by interacting with a specialized protein receptor found on the blood–brain barrier. Under conditions of very low glucose availability, neurons can utilize lactate for energy; however, as compared to glucose, lactate is not an ideal energy source for your brain. The metabolism of lactate produces much less energy than glucose—only two ATP molecules for each lactate molecule versus 38 ATP molecules for glucose. This is partly why the ancient symbiosis with mitochondria was so critical to survival.

Muscles are the primary source of lactate in the body; thus, high-intensity exercising can increase the concentration of lactate inside the muscles as well as in the blood. While the elevation in lactate is significant, it is not sufficient to alter the pH of the blood. If it did, then we would have all died immediately after any muscle activity. Your body will not tolerate changes in blood pH; even very small changes can be lethal. Blood pH is typically 7.4; if it goes up or down more than 0.5 units you will die. Due to their larger muscle and body mass (which requires more energy to move), males consistently release more lactate into the blood per unit time for a similar level of exercising than do females. Fortunately, the body's ability to remove lactate from the blood improves with continued exercise training.

A small proportion of the lactate produced while exercising is also transported into the intestines. Once inside the intestines the lactate can be metabolized by bacteria of the genus *Veillonella*. A recent study discovered that athletes whose gut contains higher levels of this bacteria have more endurance, as tested in a marathon. More recent studies have demonstrated that the conversion of lactate into the chemical propionate was the critical step that led to enhanced endurance. However, it is not currently known how increased levels of propionate in the gut lead to enhanced endurance. It might have something to do with whether whatever the bugs in your gut require propionate for their own metabolism. I have no doubt that future investigations will discover some amazing ways in which the gut microbiome influences our general health.

Elevations in blood levels of lactate following exercise may induce an increase in the release of BDNF from platelets. In the past, the assumption has been that lactate's ability to induce an elevation in blood levels of BDNF depends on a decrease in blood pH, but a recent study demonstrated that a change in pH is not required.

The presence of elevated levels of lactate in the blood induces the expression of the lactate-transporter proteins at the blood–brain barrier. Thus, the brain can transport lactate from the blood. However, the brain does not really need to, since astrocytes produce lactate and there are five times more astrocytes than neurons in the brain. Astrocytes are star-shaped cells that make up an important component of the blood–brain barrier, provide nutrients such as lactate to neurons, and carry away waste products. Thus, your neurons likely have all of the lactate they require. The brain uses lactate as a signaling molecule linking metabolism and blood flow with the increased neuronal activity required for thinking and feeling. For example, the level of lactate in the brain is temporarily elevated by visual, auditory, tactile, and motor stimuli. This temporary increase in the brain lactate level acts as a rapid source of energy for the activated neurons. Astrocytes in the brains of children with attention-deficit/hyperactivity disorder do not produce adequate amounts of lactate. Some scientists speculate that these children become hyperactive in order to increase lactate production by skeletal muscles and transport it into the brain.

### Cathepsin B

Exercising muscles also release cathepsin B. Cathepsin B is a protease—an enzyme that breaks down protein into its component parts, the amino acids. Cathepsin B may have at least one potentially undesirable function in the body because it is often found at elevated levels in cancer cells. Under normal conditions, the presence of cathepsin B often leads to cell death. Considering what is known about the role of cathepsin B in the body, its actions in the brain in response to exercise seem contradictory. Following 30 days of exercising, expression of the cathepsin B gene becomes upregulated (expressed in greater amounts). The increase in gene expression has been observed in the gastrocnemius muscle on the back of the leg as well as in the hippocampus. Exercise also increases the level of cathepsin B in the plasma. Cathepsin B can cross the blood–brain barrier. Once it is inside the brain, studies in animals have demonstrated that cathepsin B can induce the production of BDNF in the hippocampus. These findings are consistent with the idea that the production of cathepsin B by exercising muscles may be responsible for the elevation of BDNF levels in the brain. However, it is currently unknown whether, in humans, plasma cathepsin B levels and fitness or memory scores are related to each other. In addition, it is currently unknown whether the elevation of plasma cathepsin B levels is sufficient to induce BDNF production and neurogenesis in the human hippocampus in the same manner that occurs in experimental studies on animals. Confirmation of the linkage of these factors is made difficult by the fact that the possible biochemical mechanisms underlying the ability of cathepsin B to induce neurogenesis are unknown. Thus far, the findings about the potential role of cathepsin B are entirely correlational, lacking any known cause-and-effect relationship. Cathepsin B is a fascinating molecule that may play numerous beneficial and harmful roles in the brain and body.

### Lipocalin 2

The potentially harmful actions of lipocalin 2 associated with exercise are far less controversial. Lipocalin 2 is an inflammatory marker closely associated with obesity, insulin resistance, and hyperglycemia in humans. It is a good idea to try and keep lipocalin 2 levels in your body as low as possible. Thus, bodybuilders are not going to appreciate the fact that high-intensity resistance training significantly induces the release of lipocalin 2 from muscles. Why should bodybuilders care? Elevated levels of lipocalin 2 in the blood are a clinically relevant indicator of muscle and kidney damage. Lipocalin 2 levels increase rather rapidly in the blood and urine within two hours of tissue injury or exercising. Lipocalin 2 also damages the integrity of the blood–brain barrier. The elevation in lipocalin 2 levels indicates that exercising is damaging the muscles; the cellular debris produced by the muscle damage may underlie the kidney damage. When the brain detects the presence of injury associated with exercise, it releases chemicals that mimic the actions of morphine and marijuana; these chemicals reduce the sensation of pain and produce the classic "runner's high." The euphoria produced by these chemicals (more about them later) will then induce you to crave the next extreme exercising event, and the injury-induced euphoria that is so anticipated by long-distance runners. This is yet another example of the fact that your brain will often induce you to do things, such as high-intensity resistance training, that your body would prefer that you did not do.

The discovery of lipocalin 2 and the conditions under which the brain replaces pain with euphoria are consistent with the following hypothesis: Exercising is harmful to the human body, and as exercising intensifies so does the degree of damage—all of which induces the brain to release chemicals in order to nullify the pain. We humans (and every other animal that you might have seen or read about) did not evolve to exercise; we evolved to move so that we could obtain food, avoid predators, and find a mate so that our brains and bodies survive long enough to procreate the next generation. If your body is injured while performing these critical behaviors, that is an acceptable cost for the survival of you and your species. Your brain will simply instruct the body to release a bolus of pain-relieving chemicals that will also, entirely by evolutionary design, provide you with a sensation of euphoria that has the added benefit of interfering with the consolidation of the event. These ancient chemicals are little thank-you notes for fulfilling the primordial prime directive, except that this thank-you note comes attached to endogenous versions of heroin and marijuana. I will say more about these chemicals later.

## Irisin

Another important muscle-derived factor that you have probably never heard about is the protein irisin. Irisin is not a trivial hormone: Its level in blood is similar to the well-known and essential metabolic hormones insulin and leptin. Irisin is secreted by skeletal muscle following aerobic exercise, but only if the exercising lasts more than 45 minutes. The duration and the nature of the exercise play significant roles in influencing blood irisin levels. For example, high-intensity interval training and acute aerobic exercising lasting less than 45 minutes both significantly increase the level of irisin in skeletal muscle but do not produce an increase in the blood level of irisin. Apparently, irisin is produced in muscle during exercise but is not being released into the blood. Obviously, if irisin is not released into the blood, then it cannot impact brain function. In contrast, in response to a much longer duration of exercise that consumes a lot of calories, irisin is released into the blood by the muscles. The release of irisin from your muscles in response to long endurance exercising may be to inform your brain that you need to consume more calories. Consistent with this idea, irisin regulates the expression of two hormones, leptin and ghrelin, that regulate appetite. Given this role in feeding behavior, it is probably not surprising that blood levels of irisin correlate with body weight.

Irisin also influences the conversion of white adipose tissue into beige adipose tissue via its effects on the expression of mitochondrial uncoupling protein. White adipose tissue is the major site of energy storage, while beige fat releases energy as heat due to the presence of mitochondria living within the beige fat cells and due to their metabolism of fatty acids. The metabolism of fat by mitochondria is not a complete conversion into usable energy (nothing on this planet can do that, not even a nuclear reactor); some energy is lost as heat, which we refer to as calories. Scientists define one calorie as being equivalent to the energy required to raise the temperature of 1 gram (0.035 ounces) of water by 1°C (1.8°F).

It is currently unknown whether irisin can cross the blood–brain barrier and, thus, whether irisin functions as a messenger between the skeletal muscle and brain. Similar to the story for BDNF and lactate, it may not need to since the brain can make its own irisin. It is not known whether irisin released outside of the brain influences the production and release of irisin inside the brain. Acting either directly or indirectly, irisin might be a key mediator between exercising muscles and the brain. Inside the brain, according to recent animal studies, irisin may increase hippocampal neurogenesis. In order to achieve this effect on neurogenesis irisin requires assistance from another important chemical that you have probably also never heard about before, PGC-1alpha.

## PGC-1alpha

It would be hard to overstate how incredibly important PGC-1alpha is in your body. It influences the majority of metabolic pathways in every one of your cells by controlling the birth of new mitochondria. When PGC-1alpha is not functioning properly you have an increased risk of developing many unpleasant chronic diseases, including cancer. The most important feature of PGC-1alpha is that it can easily cross the blood–brain barrier. Thus, PGC-1alpha may be one of the critical messengers that muscles use to communicate with the brain.

PGC-1alpha can activate a number of antioxidant systems throughout your brain and body. The increase in antioxidant systems protects your body in multiple ways following regular exercising. Thus, although your increased muscle activity consumes a lot of oxygen, leading to increased production of toxic ROS, your brain and muscles respond by inducing mechanisms to defend themselves. Exercising elevates muscle PGC-1alpha levels, thus preventing the expression of pro-inflammatory proteins in the hippocampus. Exercise may therefore reduce harmful inflammation in the brain by inducing muscle PGC-1alpha, which then mediates the beneficial communication between skeletal muscle and the brain. Thus, with exercise, increased production of PGC-1alpha leads to increased oxidative metabolism while minimizing the negative impact of the increased oxidative metabolism (i.e., increased ROS) on cell physiology. Overall, PGC-1alpha is a wonderfully beneficial molecule to have floating around your body.

One of the consequences of not producing enough PGC-1alpha is obesity. You are already aware of the fact that brain inflammation induced by the presence of large amounts of body fat significantly impairs normal brain function. Fat cells produce inflammation by releasing proteins called *cytokines*. The fattier your fat cells become, the more cytokines get released into your blood. Essentially, obesity produces a chronic, low-grade, body-wide inflammation that reproduces many of the same metabolic conditions that underlie the aging process. Regular moderate levels of exercise, by inducing PGC-1alpha, may reduce some of the negative consequences of obesity-induced inflammation on the brain and cognitive function.

Similar to interactions involving irisin that I discussed earlier, PGC-1alpha and BDNF also interact in beneficial ways inside your brain. For example, the formation and maintenance of dendritic spines on hippocampal neurons depend on interactions between BDNF and PGC-1alpha after exercise. These dendritic spines are essential for the formation of memories. An inability to produce dendritic spines is associated with numerous cognitive disorders, including attention-deficit/hyperactivity disorder, autism, intellectual disability, and fragile X syndrome. Individuals with these disorders are generally less physically active due to many factors, such as a lack of motor coordination or the inability to maintain a regular exercise routine. Numerous studies have shown variable, but consistent, benefits of exercise in autistic children, including enhanced locomotor and manipulative skills and improved muscular strength and endurance following regular moderate physical activity. In contrast, it is currently unknown whether exercise can produce significant changes in brain physiology or chemistry in individuals with these disorders in the same way that exercise apparently influences brain physiology in individuals without these disorders.

PGC-1alpha is not always beneficial; it can also produce harmful effects in the brain and nervous system. Following a traumatic brain injury, PGC-1alpha may induce maladaptive compensatory processes that lead to spasticity, seizures, and chronic or neuropathic pain. Ironically, the same physiological mechanisms that underlie PGC-1alpha's beneficial effects— the increase in BDNF levels—may also be responsible for the processes that contribute to chronic pain and headaches related to a traumatic brain injury that most people suffer from for many years.

# 9

# THE PARADOX MOLECULE

## INTERLEUKIN-6

Every time you exercise you induce a small degree of widespread injury in the muscles and joints as well as many other soft tissues in your body. The amount of injury is, of course, related to the level and duration of the exertion. The tissue injury induces a low-level inflammatory response by your immune system. Your body's reaction to this inflammation may ultimately benefit brain and body health. For example, one of the benefits of daily exercise may be due to the counteraction of inflammation produced by daily bouts of exercise. This counteracting effect of exercising is mediated by a protein called *interleukin-6*. Interleukin-6 is a cytokine that is also a myokine (a protein released by a muscle) that plays an important role in the metabolic health of muscle cells and is released in response to moderate aerobic exercise, such as dancing, brisk walking, or gardening. This increase in interleukin-6 blood levels following moderate levels of exercise reduces the level of damaging pro-inflammatory proteins. The source of the interleukin-6 following exercise is still controversial, but it likely originates from numerous tissues and white blood cells following exercising.

The benefits of exercise exhibit an inverted-U dose–response relationship that is similar to the effects of many familiar drugs. Viewing exercise as a drug is a useful analogy: A little is good, a lot may be harmful. Interleukin-6 is a good example of this concept: It is both anti-inflammatory and pro-inflammatory, depending on the physiological circumstances that induced its release into the blood. In contrast to the benefits of moderate levels of exercise intensity, very-high-intensity exercise can trigger a much greater (up to 100-fold) increase in interleukin-6 levels. Interleukin-6 levels are greatly increased in obese people, where it stimulates a variety of pro-inflammatory and autoimmune processes that contribute to the development of diabetes, atherosclerosis, depression, rheumatoid arthritis, and Alzheimer's disease. The damaging actions of interleukin-6 in these disorders has initiated interest in developing anti–interleukin-6 drugs as therapy. A drug that could safely antagonize the actions of interleukin-6 would find widespread use. Interleukin-6 can easily cross the blood–brain barrier, where it has been shown to cause alterations in genetic function that may underlie the development of schizophrenia, another disorder of the brain that is related to chronic obesity. Elevated levels of interleukin-6 in the brain have also been implicated in the pathology of depression due to its ability to reduce the level of DBNF in the brain. In addition, interleukin-6 induces the production and release of the stress hormone cortisol into the blood. Overall, the dramatic and repeated increase in interleukin-6 associated with extreme levels of exercising can have long-term negative consequences on brain function and mental health.

Not all of your muscles release interleukin-6 in response to exercise. This fact contributes to the difficulties in understanding the role of interleukin-6 after exercise. Only oxidative muscles showed release of interleukin-6 protein; the glycolytic muscle fibers do not release interleukin-6 protein in response to a sprint or endurance run. Oxidative muscles, such as the soleus (a powerful back muscle), tend to be continuously active, weight-bearing postural muscles; in contrast, glycolytic muscles, such as the plantaris (a thin muscle in the back of lower leg that connects to the Achilles tendon), tend to fatigue rapidly. After seven days of aerobic exercising, interleukin-6 levels were reduced in both soleus and plantaris, as well as in the blood. This reduction in interleukin-6 production probably reflects skeletal muscle adaptation to continuous exercising.

### Interleukin-6 does not act alone

Recent research has found that exercising increases the blood level of a very broad range of proteins in addition to interleukin-6, which can also be either pro- or anti-inflammatory. However, only a subset of these proteins was produced by skeletal muscles in response to the exercise. Furthermore, only a few of these are secreted from the muscles in quantities that can be detected in the blood. However, they tend to be exceptionally potent molecules and many of them can easily cross the blood–brain barrier. Similar to the actions of the molecules discussed earlier, their effects in the body are complex and not always beneficial. One has been implicated in the development of psoriasis, rheumatoid arthritis, and atherosclerosis; another is critical for wound healing, cancer metastasis, and angiogenesis; and a third is associated with HIV infection. Finally, the response to exercise is biphasic: Chronic exercise induced more cytokine changes than either acute low-intensity (i.e., endurance) or high-intensity (i.e., fast sprint) exercising. Whether these other recently discovered molecules released from muscles influence the brain is currently unknown and thus are not discussed here. However, in order to understand how these molecules released from exercising muscles (and others yet to be discovered) affect your brain, I need to introduce a few critical facts about how your brain works.

# 10

# A FEW FACTS ABOUT YOUR BRAIN

Your brain, as well as the brain of your dog and the flea living on its back, is composed of cells called *neurons* that interact both electrically and chemically with each other. Your brain also contains some supporting cells that are called *glia*. One type of glia, the microglia, serves as the brain's immune system. Microglia activation is a cardinal indicator of inflammation within the brain. The inhibition of microglial activation may be one of the key ways in which exercise provides neuroprotection in the brain. The average adult human brain contains about 90 billion neurons, give or take. A typical neuron possesses a cell body, dendrites, and a long single projection called an *axon*. Dendrites are thin filaments that extend away from the cell body, often branching multiple times, giving rise to a complex tree of dendritic branches. An axon is a special narrow extension that projects away from the cell body and travels for a rather long distance. The neurons that control the muscles in the foot extend their axons for almost three feet in some humans. The cell body of a neuron frequently gives rise to multiple dendrites, but never to more than one axon. An individual neuron will communicate via its dendrites and axon with about 7,000 other neurons. Obviously, the connections are far more complicated than that of your local telephone network.

If you were to scoop out a very small section of cortex, you would find it packed with neurons, glia, blood vessels, and very little else. Most of the space between neurons is filled with astrocytes, a type of glia. Astrocytes have another more notorious role: They compose the majority of brain tumors. Astrocyte tumors are categorized as either low grade, which are usually localized and grow slowly, or high grade, which grow rapidly. Most astrocyte tumors in children are low grade; in adults, the majority are, unfortunately, high grade. Tumors that consist of only neurons are very uncommon and, when they do occur, tend to be less aggressive. The contrast between these two types of tumors, glial versus neuronal, is likely related to the fact that the normal function of glia is to multiply in number following certain types of injury or infection, while neurons are genetically programmed to never multiply once they have achieved adult status.

### Blood bringing oxygen

The brain is also densely packed with blood vessels. This is important to realize because exercise can alter the density and distribution of blood vessels, and these changes in blood to localized regions of the brain may underlie many of the benefits of exercising. Given the importance of what the brain is doing, it is not surprising that as more blood flows into the brain during exercise, the blood flow to other tissues, especially the intestines and kidneys, is significantly reduced. Blood vessels in the brain are so dense that your neurons and glia are never more than a few millimeters away from a blood vessel. A constant supply of oxygen-rich blood is critical for normal brain function whether you are exercising or not. Blood flow to the brain, about three cups per minute, is carefully regulated because too much, associated with high blood pressure, or too little blood flow, associated with ischemia or stroke, can be harmful to the brain. Normal brain function and consciousness are rapidly lost if the supply of oxygen is interrupted for even a few moments. The regulation of the blood flow in the brain is fundamental for all neural activities.

Many of the immediate benefits of moderate levels of exercise on brain function are directly related to increased blood flow into the brain. After moderate levels of exercise most people feel as though they can think more clearly, their mood is modestly elevated, and they fall asleep faster. Blood flow to the brain increases incrementally as the exercise load to the muscles increases—but only up to a point. One study used an advanced magnetic resonance imaging (MRI) technique to measure cerebral blood flow at different levels of exercising. They discovered that cerebral blood flow decreased significantly throughout the brain in patients who participated in fatiguing aerobic exercise or tasks with a high physical load. Overall, blood flow into the brain increases in a rather linear fashion as the level of exercising increases from mild to moderate levels. However, with extreme levels of exercising, blood flow into the brain is decreased. The brain benefits the most from participating in mild to moderate levels of exercise.

## The blood–brain barrier

The astrocytes carefully control what is able to cross from blood to brain; they are a critical component of your "blood–brain barrier." The blood–brain barrier permits the easy entry of only a few substances into the brain. Fat-soluble substances can easily enter the brain. Very small molecules, particularly if they do not carry an electric charge, usually get through the blood–brain barrier. The brain actively imports the nutrients it requires from your diet through the blood–brain barrier. Some regions of the brain lack any blood–brain barrier. The barrier does not form in these brain regions so that your brain can monitor levels of specific chemicals, such as the presence of sugar in the blood. Your brain is only capable of sensing the presence of sugar, actually glucose, in the blood; it cannot sense the levels of proteins or fats in the blood. In addition, because of the brain's unique blood–brain barrier feature and its inability to store excess fuel in any form, the brain requires a continuous supply of glucose to sustain all cognitive activities. It has been estimated that an average brain consumes about one full cup of glucose, about 215 grams, each day! This is about twelve regular-sized donuts; now you know why you love them so much. The brain accounts for about two-thirds of the body's total glucose consumption during rest. Thus, a meal devoid of carbohydrates often leads to a feeling of constant hunger, dizziness, irritability, and difficulty concentrating or speaking. Your brain hates low-carbohydrate diets such as the ketogenic diet. Indeed, study after study has confirmed that the ketogenic diet does not enhance any measure of brain function, including mood, sleep, attentional abilities, or executive functions. Overall, the ketogenic diet is based on completely flawed pseudoscientific nonsense.

### The ocean in your head

The brain is also protected by three layers of protective membranes called *meninges*; their names are the *dura* (the outermost layer), *arachnoid* (the middle layer that resembles a spider web), and *pia mater* (the innermost layer). A person has meningitis when these membranes become infected. Between the arachnoid and pia layers is a space filled with a clear, colorless liquid. This liquid is cerebrospinal fluid and is essentially blood that has been filtered of cells and most proteins. Freshly produced cerebrospinal fluid is constantly rinsing your brain. The amount of fluid produced is impressive: Every ounce of cerebrospinal fluid is completely replaced about four times every day. If the constant flow of cerebrospinal fluid is impeded in any way, the fluid will quickly accumulate inside your skull, increasing intracranial pressure, pressing the brain against the inside of the skull, and squeezing the little blood vessels that feed the brain until they close, ultimately leading to the death of brain tissue. This condition is called *hydrocephalus*; it occurs more often in infants than older adults and can be fatal if not corrected immediately. Your skull has a fixed volume; this fact often places your brain at serious peril.

The brain is submerged in an ocean of cerebrospinal fluid inside your skull. Why go to all of this trouble to keep the brain afloat? The answer is related to the fact that your big brain weighs a lot: about three pounds. However, when floating in the cerebrospinal fluid the net weight of your brain is equivalent to a mass of only 25 grams—that is less than an ounce of beer. How is this possible? If you have ever experienced floating in the ocean, the salt water made you buoyant and you floated easily on the surface with very little effort, as though you weighed much less. The salty cerebrospinal fluid provides the same benefit; it allows the brain to maintain its density and shape without being crushed by its own weight. This same principle of buoyancy allows seagoing mammals, such as whales, to become very large; however, once they have left the salt-water ocean, they quickly succumb to the consequences of their own weight. If your brain were resting on a table, its own weight would quickly crush the small blood vessels supplying it, killing the cells on the bottom of the brain.

Sometimes, depending on the type of exercise you are performing, the compartment containing the cerebrospinal fluid may develop a small leak that causes very painful headaches. Neurologists call these *low-pressure* headaches. The headache is caused by a decline in the amount of cerebrospinal fluid that your brain is floating on, leading to a slight distortion of the dura. The dura has pain sensors; your brain does not. Low-pressure headaches usually occur after participation in exercises that involve twisting movements of the body, such as tennis, canoeing, fly fishing, yoga, or weightlifting. Low-pressure headaches are usually associated with pain that radiates across the back of the head as well as in the neck and upper shoulders. Due to the nature of what is causing the pain, most people report that the pain is made worse by coughing or standing and is reduced by lying down.

### Blows to the brain

In addition, by floating within the skull, bathed in cerebro-spinal fluid, the brain is protected from being injured when the head is quickly jolted around. Unfortunately, when the head is violently displaced due to a blow to the head associated with boxing or playing football, soccer, or field hockey, this fluid buffer cannot prevent the brain from colliding with the inner surface of the skull. If the brain does collide against the skull, the outer layer of the brain, the cortex, can become attached by scar tissue to the inner surface of the skull, leading to the death of outer layers of the cortex. Repeated violent blows to the head, such as those experienced by boxers and other athletes with a history of repetitive brain trauma, ultimately lead to large sections of the cortex becoming stuck to the inner surface of the skull. (I once performed an autopsy on a boxer. It was nearly impossible for me to insert my hands between his brain and skull because the outer layers of his cortex had fully attached to the inside of his skull.) If the injury to the cortex continues and becomes widespread, the resulting loss in brain function is called *dementia pugilistica*; in the 1920s these symptoms were called the *punch-drunk syndrome*. Today, due to the notoriety drawn to this condition by professional athletes, this progressive degenerative disease is referred to a *chronic traumatic encephalopathy*. The damage associated with this injury to the cortex may spread to involve nearby subcortical brain areas that are responsible for movement and sensory processing, producing symptoms such as stiffness, slowness, and walking or balance problems; when this happens, the result is Parkinson's pugilistica. The boxer Muhammad Ali demonstrated many of the symptoms of both dementia pugilistica and Parkinson's pugilistica.

### The basic brain plan

Within the brain, most of the major structures evolved as small clumps of cells, called *nuclei* or *ganglia*, which are involved in related functions. Some ganglia control movement, some control body temperature, and some control mood. Overall, the basic plan, whether you are an octopus or a human, is that neurons communicate with one another in order to facilitate the sensing of the external world and the internal events taking place inside the body. The brain then decides which behavior to elicit in order to improve its chances of survival and the propagation of its species. For humans there is sometimes a more ephemeral goal: Do something that brings pleasure. For humans during the 1960s, this was succinctly re-envisioned as sex, drugs, and rock and roll. For many humans today, pleasure is also attained with regular exercising. We experience pleasure thanks to the particular neurotransmitter chemicals that are being released in the brain and to the particular regions of the brain where they are being released.

The function of each neurotransmitter depends entirely on the function of the structure in which it is located. Let us look at a few examples. Deep within your brain is a region called the *basal ganglia*. The neurons in the basal ganglia are responsible for producing normal well-controlled smooth movements. The level of the neurotransmitter dopamine in these nuclei is much higher than in most surrounding brain regions. Therefore, scientists have concluded that dopamine within the basal ganglia is critically involved in the control of movement. Furthermore, if we expose your brain to a drug that impairs the function of dopamine, then your ability to move will be impaired. Dopamine is obviously critical for movement. It would be incorrect, however, to assume that dopamine is *only* involved with the control of movement. You also can find dopamine in the retina of your eye and in your hypothalamus, structures that have nothing to do with movement. Dopamine also is released into small regions deep within the frontal lobes; when this happens, you experience a feeling of pleasure. The takeaway point from these examples is that there is no such thing as a specifically unique "dopamine function" or an exclusively distinct "serotonin (another neurotransmitter) function." The brain region within which the neurotransmitter is found defines its function, not the neurotransmitter itself. Many neurotransmitters other than dopamine are associated with the experience of pleasure; for example, the release of acetylcholine in the septal area produces a feeling a well-being and joy, while the release of enkephalin or anandamide within the cortex produces a feeling of euphoria. It is both naïve and incorrect to refer to dopamine as the brain's only "feel good" neurotransmitter; it is not.

### Brain chemicals

Neurotransmitter chemicals are produced from the contents of your diet: Thus, what you eat can, under certain conditions, influence how you think and feel. First, nutrients such as amino acids, sugar, and fats are absorbed from your food and transported across the blood–brain barrier into your brain. These nutrients are then absorbed into your neurons, where specialized enzymes convert them into neurotransmitters. The neurotransmitter molecules are then stored in very tiny spheres, which sit patiently waiting for the arrival of an action potential that instructs the neuron to release them from its terminal. Once outside the neuron, the neurotransmitter wanders around looking for a way to communicate with the next neuron. The junction at which two neurons communicate is called a *synapse*. The neurotransmitter molecule, now free to wander around within the synapse (which is in fact a very small space), will ultimately bump into and connect with a special protein called a *receptor*. Receptors are like boats floating on the outer surface of the neuron on the other side of the synapse. Receptors offer comfortable docking ports for the neurotransmitter to attach itself.

Once this docking of neurotransmitter and receptor has been achieved, the next stage in the communication process between neurons begins. At this point, lots of different things could occur in response to the neurotransmitter binding to a receptor; ions might move in or out of pores, enzymes might be activated, genes might be turned on or off, and many other possibilities. These secondary processes can have long-term consequences for the neuron's behavior and ultimately for your thoughts and actions.

Meanwhile, back in the synapse, after interacting with the receptor, the actions of the neurotransmitter must be terminated by means of its reabsorption back into the neuron that originally released it. This vacuuming-up process is called *re-uptake*. Later you will learn about drugs that block this reuptake process and their role in treating depression and anxiety far more effectively than exercising. Alternatively, the neurotransmitter also might be acted on by local enzymes and converted into a chemical that can no longer interact with your brain. Once the neurotransmitter is inactivated, it is removed from the brain into the bloodstream. Such byproducts of the ordinary hustle and bustle of the brain can be monitored easily in many of our body fluids, and this information can be used to determine whether our brains are functioning normally. One thing that does not happen: Neurotransmitters produced in the brain do not leave the brain intact. They either are metabolized or their escape is blocked by the blood–brain barrier. This blockade is crucial because if these neurotransmitters escaped from the brain, they might have lethal or very unpleasant consequences. The reverse is also true: Neurotransmitters produced outside the brain cannot get in. The ability of muscles to communicate with the brain is often prevented by the blood–brain barrier.

## Instructing your muscles

Inducing a muscle to contract involves processes that are both electrical (via the passage of an electrical disturbance called an *action potential* traveling down the axon) and chemical (via the release of a neurotransmitter onto the muscle). An electrical signal is sent out from the brain toward a muscle waiting at the end of the axon. The arrival of an electrical signal induces the release of a neurotransmitter chemical called *acetylcholine*. Acetylcholine binds to specialized protein receptors that induce the contraction of the muscle. Sometimes this process is compared to activity in an old-fashioned telephone system. Imagine that when your phone rings, the electrical signal traveling via the telephone lines outside your house that brought the call into your telephone is similar to an action potential traveling down an axon. Someone has sent you a signal. Now imagine that you pick up the telephone receiver, you hold it to your ear, and the phone spits some chemicals into your ear. Your ear is the receptor for the chemical. Let us imagine that your ear is connected directly to a muscle. The communication works the same way whether individual neurons are communicating with each other or whether the brain wants to instruct a muscle to contract. Everywhere in your brain and body, neurons are being electrically induced to spit out chemicals. Ultimately, a muscle contracts and your body moves.

### How the parts work together

Now that you are familiar with the individual components of the brain, neurons and glia, and the neurotransmitters that they use to communicate with each other, let us put all of the pieces together and examine how the brain is organized to control movement. During the past few decades, with the introduction of noninvasive techniques to examine brain function, neuroscientists have resurrected, rather inadvertently, a description of brain function that resembles a discredited idea from 1796. In that year a German physician, Franz Gall, developed an approach to understanding the brain by focusing on measurements of the human skull based on the concept that certain brain areas, or modules, have specific localized functions. The idea was intuitively attractive and became quite popular; in addition, the approach rather crudely described how the brain actually functioned. Today, over 200 years later, thanks to the invention of some highly sophisticated and expensive scanning machines, we have returned to the concept of a compartmentalized brain. Some parts of the brain are undoubtedly dedicated to specific functions, such as vision, hearing, or touch; thus, the 200-year-old idea of brain modules is not entirely erroneous. A better analogy, however, is to view the brain as an orchestra that requires many unique instruments (brain regions) to work together simultaneously to produce a complex pattern of activity leading to the emergence of a beautiful piece of music (e.g., a perception of a bluebird flying across one's field of vision on a sunny day). Please keep this concept of an orchestra in mind as we introduce the functions of various brain regions, beginning with what happens in the front half of the brain, where movement is controlled.

### The frontal lobes

Just behind each eye, in the front half of your brain, are the frontal lobes. The frontal lobes do our thinking and control movement in response to those thoughts. Much of what is known about the function of these frontal lobes was learned by examining the behavior of people who suffered traumatic brain injury, stroke, cancer, or infection within that area of the brain. During the past few decades the use of noninvasive technologies, such as the MRI scanner, has provided the opportunity to monitor the activity of the frontal lobes in conscious humans performing specific tasks. This is what scientists have learned.

The frontal lobes allow you to make decisions, plan your actions, organize your thoughts about specific goals, and then instruct your body to move in response to your goals. Once your frontal lobes have made a decision, there is only one thing that they can do—instruct some muscles to contract and move a part of your body. Despite the complexity of our frontal lobes and the sophistication of their neural processes, our brain has a very limited number of options—it instructs a muscle to contract to move a limb or finger to pick up a second piece of cake or type the letter "a." That's about all it can do in response to our wonderfully complex thoughts—move you from here to there, and back again.

Your frontal lobes are responsible for expressing the many personality traits that you inherited from your parents. It is probably not too surprising to learn that variances in the size of some regions of your frontal lobes correlate with specific personality features. For example, introspection is correlated with the size of a region that lies on the top part of the frontal lobes called the *prefrontal cortex*. If you move your focus to a region just lateral to the part of the frontal lobe (just behind your eyebrows) responsible for introspection, you will find a region that becomes active when we perform complex behaviors such as paying attention or lying. Yes, your brain contains a region that appears to be devoted to lying.

Lying is apparently a complex task that requires considerable attentional ability, a vast memory for past events, and the significant participation by this frontal brain region. These highly evolved brain regions allow humans to be exceptionally good liars. Psychologists believe that most of us tell a lie to someone we know at least twice a day, and that within a period of one week, we lie to nearly one-third of the individuals we meet. Some individuals, such as narcissists, are born with cortical proclivities that induce them to lie almost constantly for the simple reason that they do not care about telling the truth. Narcissists often succeed as businessmen and politicians.

If you now move your focus to the most lateral aspect of the frontal lobes, just inside of your temples and behind your eyes, you will find a region, the inferior frontal cortex, that is responsible for controlling low-risk behaviors. This region is in constant communication, and competition, with a region located deep with the brain called the *nucleus accumbens* that induces you to participate in high-risk behaviors. These two regions compete for control when you are trying to decide whether to have another piece of chocolate cake: The inferior frontal cortex is saying, "No, you do not need another piece; it will just make you fat," while the nucleus accumbens is saying, "Eat it! It will taste so good!" Once your frontal lobes have made a decision, there is only one thing they can do—instruct some muscles to contract in order to scoop up that last piece of cake.

## The back half of your brain

Across the many different species that have been studied, the locations of specific functions have remained surprisingly quite consistent; for example, motor functions are always controlled by the frontal lobes, while sensory information coming into the brain is processed by the back half of the brain. Overall, your brain, as well as the brain of your cat and the mouse it is waiting to catch, is organized so that the back half receives incoming sensory information and then processes it into your own very personal experience of the here and now. The lobes in the back half of your brain process the voice that you hear, smell the aroma of food cooking, feel a craving for food as your blood sugar levels fall, sense that it's late in the day and the sun is setting and the room is getting darker, and thus, it must be dinnertime. Once sights, sounds, smells, taste, and feel of life events are processed by the back half of the human brain, this information is then funneled into the temporal lobe, located just inside where your ears attach to your head, where it becomes organized for long-term storage. A critical part of this process requires the hippocampus. In order to understand how exercise improves brain function, you will need to understand the role of the hippocampus.

## The hippocampus and neurogenesis

The hippocampus is located within the temporal lobe and is responsible for making memories. The brain continually rewires itself as you experience life and learn new things. This ability to make new connections to form new memories is termed *neural plasticity*. Plasticity allows you to be smart, adaptive, and better able to survive in a changing and challenging environment. The effects of exercise on plasticity will be discussed later. Plasticity also has a dark side; it underlies your tendency to become addicted to drugs, foods, sex, gambling, and potentially dangerous or risky behaviors. The hippocampus also has another very special feature that is shared by no other region of the brain: It contains a cellular nursery that gives birth to new neurons every day. It was once believed that all of your neurons were born while you were developing inside the uterus and that only a small percentage were added during the first few years of life. The assumption had always been that adult brains possessed all of the neurons that anyone would ever need. Apparently, this is not true; you actually require a constant supply of new neurons when you are an adult too. Adult human brains give birth to approximately 700 new neurons every day. That is quite a lot of new cells to assist with neural processing. This process of neuronal birth is called *neurogenesis* and occurs within the hippocampus in humans. Some other species also maintain neurogenesis within their olfactory system; humans do not. It was once believed that hippocampal neurogenesis continued throughout adulthood. This is no longer thought to be true. As humans get older, neurogenesis declines precipitously. Chemicals released by exercising muscles, and some medications, may significantly induce adult hippocampal neurogenesis. When your muscles do not function correctly due to disease or a neurological problem, they can have a negative impact on brain function. Clinical evidence confirms the importance of normal muscle activity; for example, children with Duchenne muscular dystrophy have cognitive deficits characterized by learning disabilities. Mouse models of human muscular dystrophy also demonstrate impaired memory function. Clearly, healthy muscle activity underlies, either directly or indirectly, healthy brain function.

The many benefits of exercise that are likely due to increased neurogenesis include reduced levels of anxiety and depression, enhanced social skills and self-esteem, as well as improved mood and cognitive abilities. The discovery of neurogenesis was an important scientific breakthrough that significantly changed how we understood the nature of brain function. The effects of exercise on neurogenesis will become clear in the following chapters.

### The little brain and movement

Our journey inside your skull concludes with an examination of the function of the little lump of brain in the back. Long ago, anatomists saw what appeared to be an additional companion brain hanging off the back of the brain and decided to call it the cerebellum, or "little brain." It is a tennis ball–sized structure, only about a tenth of the size of the brain that, surprisingly, is densely packed with almost 50% of all of the neurons in your head. It has a highly convoluted outer cortical shell that has only three layers in contrast to the six layers of your brain's cortex. The interior of the cerebellum contains lots of axons going to and coming from the brain and spinal cord.

What does the cerebellum do? Once again, the answer to this question comes from studies using noninvasive scanning machines and indwelling electrodes. The cerebellum plays a role in the control of certain types of memory and mood. Given the importance of the cerebellum for coordinated complex movements, particularly those most often associated with exercise, you might predict that acute aerobic exercise would have some effect on this function of this brain structure. However, one recent study provided only modest, nonsignificant, evidence for changes in neural circuits within the cerebellum in response to regular exercise.

Neurons within the cerebellum become active just prior to and during the activation of the muscles of the body. When the cerebellum is damaged due to injury, stroke, or tumors, the most common symptom is difficulty with movement and posture. People with cerebellar damage can move, but their movement is not smooth or well controlled. The cerebellum receives sensory information from your muscles and joints to inform you about the location of your body parts; this allows you to move correctly without paying close attention to every movement. Usually patients with cerebellar damage find that they need to walk with their feet placed widely apart and, because they cannot tell where their limbs are located, they need to watch what their limbs are doing at all times.

### Alcohol and movement

The ability of the cerebellum to perform complex well-learned movements smoothly can also be impaired by alcohol intoxication; this is because the cerebellum contains specific receptor proteins that alcohol binds to and as a consequence interferes with normal function. If you are ever stopped by a police officer on suspicion of driving drunk, you may be asked to touch your nose with your outstretched finger. Ordinarily this is quite an easy task to perform; however, it is not easy to perform when the cerebellum is bathed in alcohol. Alcohol distorts the pattern of neuronal activity in your cerebellum, preventing you from moving your arm accurately to touch your nose. Alcohol also distorts the ability of your cerebellum to control the smooth coordination of the muscles of the eyes. When the police officer instructs you to follow his finger with your eyes, he is testing to determine whether alcohol has impaired the ability of your cerebellum to control the muscles of your eye. If your eyes begin to move involuntarily from side to side in a rapid, swinging motion rather than staying fixed on the officer's finger, you are displaying what is called *nystagmus*. Nystagmus can be induced by alcohol intoxication.

### Connecting to your muscles

Pathways of nerves that begin within the front half of your brain run throughout your body to make connections with each muscle. The terminal end of the axon projection that connects to a muscle releases the neurotransmitter acetylcholine, which induces the muscle to contract. The precise mechanism that allows acetylcholine to contract a muscle is beyond the scope of this book. Acetylcholine plays many roles in the body in addition to inducing muscles to contract. Exercising requires the production and release of quite a lot of acetylcholine every minute of every day and night (you are contracting muscles even while you are sleeping). Your body requires a continual supply of the dietary ingredients for the production of acetylcholine. Where does your brain obtain these critical nutrients that underlie thinking and movement? Here is one potential source.

## Coffee and donuts

Before you get to the point in your drive to work this morning when you stop for your daily cup of coffee and a donut, let us back up a couple hours to when you were sleeping. The alarm rings, you awaken, and you are still drowsy. Why? Being sleepy in the morning does not make any sense; after all, you have just been asleep for the past eight hours. Shouldn't you wake up refreshed, aroused, and attentive? No, and this is why. During the previous few hours before waking in the morning, you have spent most of your time dreaming. Your brain was very active during dreaming and quickly consumed large quantities of the energy molecule ATP. The "A" in ATP stands for adenosine. The production and release of adenosine in your brain are linked to metabolic activity while you are sleeping. There is a direct correlation between increasing levels of adenosine in your brain and increasing levels of drowsiness. Why? Adenosine is a neurotransmitter that inhibits (turns off) the activity of neurons responsible for making you aroused and attentive. You wake up drowsy because of the adenosine debris that collected within your brain while you were dreaming. So, what does your brain need so that you will feel better and pay attention to the traffic? Before considering the answer, please keep in mind that what your brain wants is not always good for your body. You can now appreciate why it is so hard to attain a truly complete mind–body balance. Your brain evolved mechanisms and behaviors that enhance your survival until you have procreated. Appreciating that fact makes the next few paragraphs easier to understand.

Your brain is not cooperating when you wake up because it lacks one critical ingredient that it urgently needs: sugar. You have been fasting since dinner last night and your blood levels of sugar have fallen. From your brain's perspective, sugar is absolutely indispensable. It will do whatever is necessary to convince you to eat sugar as often as possible. Why? Your brain needs sugar (usually in the form of glucose) to function normally. The billions and billions of neurons in your brain require a constant supply of sugar to maintain their ability to produce energy and communicate with other neurons. Your neurons can only tolerate a total deprivation of sugar for a few minutes before they begin to die. Therefore, as blood levels of sugar decrease with the passage of time since your last meal, you begin to experience a craving for food, preferably something sweet. Essentially, the presence of sugar in your brain is considered normal, and its absence leads to the feeling of craving and the initiation of foraging behaviors, such as seeking out a candy bar or a donut. There is a reason that donut shops and sugar-laden cereals are so popular in the morning (or, for some people, all day long), and you can lay the blame on neurons within the feeding center of your hypothalamus. If your brain did not want those donuts so badly, the donut shops would not be so densely distributed along your route to work.

Once inside the brain, the sugar in those donuts you just ate is used to produce acetylcholine. In addition to being critical for movement, acetylcholine allows you to learn and remember, as well as to pay attention to traffic. Your brain also requires choline to make acetylcholine, which is also obtained from the diet. We frequently obtain choline in our diet by eating lecithin. Lecithin can be found in many different bakery goods such as donuts and cupcakes and is commonly added to chocolate. Thus, a tasty chocolate-covered donut first thing in the morning is going to provide your brain with the sugar and choline it needs to produce acetylcholine in order for you to pay attention, learn new things, and contract your muscles.

As the day progresses, the millions of acetylcholine neurons in your brain and body are busy contracting muscles and consuming choline and sugar as you spend your day thinking and learning. This is why your brain utilizes the equivalent of 12 donuts of sugar every day. Now, as evening arrives, you notice that you're having trouble paying attention and you're experiencing some mental slowing. What's happening in your brain and what can you do about it? The cure for your mental slowing: coffee. While you were busy thinking and learning all day, another neurotransmitter chemical was increasing in concentration, and it has slowly and powerfully begun to turn off your acetylcholine neurons. The culprit is once again adenosine, the "A" from ATP. All of that thinking and moving burned up a lot of ATP that was produced by the metabolism of fat and sugar in your brain. Adenosine once again is actively inhibiting the function of acetylcholine neurons in your brain; the longer you are awake, the more persuasive is the inhibition. Fortunately, there is an easy solution to your dilemma. Caffeine in coffee and tea is able to block the actions of adenosine and release your acetylcholine neurons from their chemical shackles; your attentiveness improves and you are ready for anything—at least until the caffeine effect wears off.

Now you can easily understand why coffee and donuts are so often sold together. The owner of your local donut shop is obviously a very knowledgeable neurochemist. Your brain and muscles need them to be there every morning on your way to work. That is why you see so many coffee and donut shops near where you live.

## Where "YOU" live

Let us consider where YOU live. In this case, I am not referring to where your house is located; I am referring to the location inside your brain where your consciousness lives. Thinking involves the release of neurotransmitters, such as acetylcholine, from the end of one neuron on to the next neuron and so on. The neurons in the frontal lobe release lots of different neurotransmitters in order for you to think and plan movements such as dancing and speaking. YOU live in the frontal lobe of your brain. Yes, the YOU who is thinking about the meaning of the last sentence; the YOU who is feeling hot or cold right now, or hungry, or angry, or absolutely anything at all—that YOU. This particular region of cortex is part of a small circuit of brain systems that turn on as soon as you wake up in the morning. Think of that very familiar feeling you have immediately on awakening that you know who you are and where you are in relation to familiar objects and other people. The frontal lobes are also one of the brain regions that become active when your mind is wandering or when you are simply recalling episodes from your past or contemplating actions in the future. Thanks to recent studies we now understand that the region of cortex on the medial (deep inside between both hemispheres) surface of the frontal lobes is responsible for allowing us to have a "Theory of Mind," or the ability to recognize that other people have thoughts and feelings that are independent of ours. Essentially, YOU live about five centimeters behind the center of your forehead.

Now that you know where YOU live in your brain and how your brain decides to move your body by contracting your muscles, let us assume that you have decided to exercise by repeatedly contracting muscles until your brain is informed by your incoming sensory neurons that its body is becoming hot, sweaty, and tired and that your heart is beating quickly and your lungs are expanding and contracting rapidly.

## Enhancing dendrites

The act of exercising, or doing almost anything at all, induces changes in how individual neurons in your brain talk to each other. This process involves many different types of biological processes; those at the level of the DNA occur within a few hours, while those at the level of the synapses may take many weeks to develop. Neurons communicate with each other by releasing chemicals onto specialized structures called *dendrites*. Dendrites form a tree-like set of branches that allow for thousands of communication points between neurons. Some neurons connect to the next neuron via specialized nubs on the surface of dendrites that are called *spines.* These spines are specialized types of synapses that enhance the communication between neurons. Exercising increases the number of dendritic spines and enhances the complexity of the dendritic tree. Most of these dendritic enhancements have been documented to occur within a brain region, the temporal lobe, which is critical for learning and memory as well as depression and anxiety. In addition to enhancing the dendrites, exercise also enhances the birth of new neurons, called *neurogenesis.* It is generally believed that neurogenesis is beneficial to the brain; however, no one has figured out the specific details about how it is beneficial. Most of the evidence is still correlational. The newly born cells may integrate themselves into existing neuronal circuits after they mature. This explanation assumes that these new neurons replace existing mature neurons within these circuits that have become dysfunctional. Apparently, the brain contains quite a few dysfunctional neurons that need daily replacement given that the hippocampus generates almost 1,000 new neurons every day. However, not all of these new neurons become inserted into existing brain circuits; many of them will likely be instructed to commit cellular suicide, called *apoptosis*. Movement has always been a required part of survival, and neurogenesis into adulthood may enhance our ability to survive.

It would be advantageous if movement and neurogenesis were somehow functionally linked to each other. As you will read in the following chapters, they are.

# 11

# HOW YOUR BRAIN RESPONDS TO EXERCISE

*Getting the balance right*

Overall, different types of exercise yield different consequences on brain function. For example, aerobic exercise increases blood flow into the brain, increases oxygen consumption, improves reaction time, and enhances performance in tests of executive functioning; in contrast, anaerobic exercise does not produce equivalent benefits. The effects of exercise on brain function can be compared to the effects of drugs on brain function. Similar to the effects of drugs, the effects of exercise can be described by an inverted-U–shaped curve; too much or too little exercise is not advantageous and may even be harmful. This is another example of the principle of moderation that I mentioned earlier: Everything must be just right; not too much, not too little. For example, intense exercise greatly reduces blood flow to the brain, impairs spatial memory acquisition, produces considerable mental fatigue, and leads to an excessive production of harmful ROS that may overwhelm the brain and body's natural antioxidant defense systems. Too little, or infrequent, exercise will fail to upregulate the body's antioxidant capacity enough to counteract the effects of exercise-generated ROS. In contrast, regular exercise of moderate intensity can substantially improve cognitive function by reducing the effects of harmful ROS. Optimal levels of daily moderate exercising increases the levels of neurotrophic factors, reduces oxidative stress, curtails neuroinflammation, improves cerebral blood flow, or some combination of these.

So what is a moderate exercise? Generally, in order to make this judgment, people pay attention to how the exercise affects their heart rate and breathing. The "talk test" is an easy way to measure relative intensity: If you can talk but not sing, you are probably doing a moderate level of exercise. The precise influence of intensity, time, and duration on the benefits of different types of exercise has not been convincingly defined. Some studies reported that moderate intensity and duration of exercise correlate with better outcomes as opposed to either mild or exhausting exercise. Thus, there does appear to be an optimal level of exercise that produces the most benefits for the brain. Moderate levels of exercise intensity promotes long-term metabolic and hormonal adaptation and induces beneficial effects on the brain. Long-term moderate exercising also significantly reduced blood levels of pro-inflammatory proteins as well as C-reactive protein. In contrast, performing just a single bout of exercise was not very helpful. Worse, a single bout of exercise enhanced the level of pro-inflammatory proteins in patients diagnosed with diseases associated with inflammation, including type 1 diabetes mellitus, cystic fibrosis, and chronic obstructive pulmonary disease. C-reactive protein levels are also increased in the hippocampus of patients with neurodegenerative diseases, such as Alzheimer's disease. Long-term inflammation in the brain places the brain at risk of developing many different degenerative illnesses. In general, studies on humans have reported that regular moderate levels of exercise reduce C-reactive protein expression. In contrast, high-intensity exercise does not.

### Too much can be too much

The available evidence provides support for both beneficial and deleterious influences of exercise on brain function. This is the essence of the principle of moderation that defines the behavior of so many different biological processes associated with the human body. I have already listed lots of examples. Too little oxygen and you die; too much and you age your body too fast. Too little food and you die; too much and you induce inflammation, which also ages your body too fast. A little inflammation defends your tissues; too much inflammation damages them. Exercise also requires a balance between too much and too little.

One important factor is the intensity of physical activity. Studies have demonstrated a relationship between exhausting exercise and increased generation of harmful ROS. This matters quite a bit with regard to your brain because it is much more vulnerable to the harmful effects of ROS as compared to the rest of the body. Brains are more vulnerable than your other organs because they possess low levels of endogenous antioxidant defenses and high levels of unsaturated membrane lipids that can be converted by ROS into toxins. The ROS generated by exhausting or extreme exercising can induce the simple amino acids floating around inside of your body to switch from being beneficial to harmful. The ROS generated by the body can easily cross the blood–brain barrier. Once inside the brain, the ROS induce the neurotransmitters dopamine and serotonin, which are both involved in controlling your mood, to transform into toxins that are, ironically, lethal to dopamine and serotonin neurons. This chemical process is nearly identical to what happens in the brain of someone who uses the street drug Ecstasy. Numerous studies, some conducted in my laboratory, on the chemical cascades set in motion in the brains of Ecstasy users have led neuroscientists to conclude that the sustained production of ROS also likely underlies the pathophysiology of many age-related degenerative diseases of the brain, particularly Parkinson's and Alzheimer's disease. Taken together, the available evidence strongly suggests that high-intensity exercising induces changes in the brain that resemble those produced by both aging and illicit drugs. That evidence alone should convince someone to avoid high-intensity exercising. If not, then keep reading.

High-intensity exercise also increases the level of pro-inflammatory proteins, called cytokines, in the hippocampus. The elevation of these pro-inflammatory cytokines underlies the loss of neurogenesis in the hippocampus following excessive exercising. Recall that neurogenesis is critical for normal learning and memory abilities and may be necessary to avoid becoming depressed. Taken together, these studies provide insight into the changes in brain chemistry that explain how excessive physical activity conducted over a long period of time can produce psychological symptoms that resemble general cognitive decline and clinical depression. Good health requires maintaining an optimal balance between how much you move and the generation of the energy required to move. As I discussed in detail earlier, the biological rules that govern this balance between energy production, energy utilization, and the critical role of oxygen were negotiated about 3.5 billion years ago. This primordial arrangement cannot be renegotiated.

As a consequence, the brain evolved biochemical mechanisms to defend itself because it generates ROS as a consequence of doing all of the amazing things that brains do. However, the brain can only defend itself up to a certain level. When the production of the ROS exceeds the ability of the brain's endogenous antioxidant systems to eliminate them, oxidative damage occurs and neurons die. Your brain is generating ROS while you are reading and thinking about this sentence. (Please do not stop reading and thinking!) Thus, your brain is exposed to ROS every moment of every day throughout your life. It is unavoidable and underlies why every day of your life your body and your brain get a little older. Fortunately, there is a simple partial solution that involves exercising. Moderate daily exercise preconditions the brain to better adapt to the consequences of ROS production. This is one of the potential benefits of daily moderate exercise to the brain: Exercise helps your brain to tolerate the consequences of its own functioning.

The relationship between increased neural activity and ROS production raises some interesting issues related to the high cost of having such a big brain in a body with so many muscles. Large brains are metabolically expensive to maintain and protect from the consequences of their own production of ATP. Exercise places the brain at added risk from oxidative damage due to greatly increased levels of mitochondrial activity. Swimming and running have been shown to significantly increase the level of ROS in the brain. These discoveries predict that, over the course of a lifetime, the formation of too many ROS due to too much physical exercise may negatively affect brain health.

In contrast, moderate exercise leads to physiological adaptations that increase the production of a healthy balance of pro- and anti-inflammatory cytokines in the body; these molecules can easily enter the brain and indirectly support neurogenesis. In contrast, high-intensity exercise leads to the accumulation of oxidative damage and a significant increase in the activation of ROS-generating processes in the brain. Thus, overall, the differential effects of moderate exercise as compared to intense exercise may be explained by the differential generation of ROS by mitochondria. Moderate exercise requires less oxygen and produces fewer ROS; intense exercise requires more oxygen and subsequently generates more ROS.

Moderate levels of ROS associated with moderate levels of exercise may be necessary for healthy communication between neurons. Rather paradoxically, your brain requires an "optimal" level of ROS production for normal brain function. This may explain why consuming excessively high doses of antioxidant supplements impairs hippocampal neurogenesis: They may reduce ROS production too far. Maintaining a balance of these factors over a lifetime is critical. The moderate levels of ROS produced in the brain by moderate levels of exercise also elevate BDNF levels, which can, paradoxically, reduce ROS production.

Catalase is a critical antioxidant in the brain that it is capable of reducing the level of ROS. One of its important functions is the detoxification of the ROS produced as byproducts of the oxidation of dopamine that I referred to earlier. Oxidation is, as the name suggests, due to the actions of oxygen molecules floating around in your body following the increased consumption of oxygen associated with respiration. The level of these dopamine oxidation byproducts is increased after intense exercising. This is yet another example of the ironies associated with exercise. Dopamine is required for the control and coordination of complex movements, yet it becomes vulnerable to the consequence of too much movement. Catalase can defend your dopamine neurons in response to exercising, but only up to a certain point. Good brain health requires maintaining a good balance between how much you move and the generation of the energy your body requires to move.

These discoveries lead to an interesting prediction about movement, energy production, breathing in oxygen, and the death of dopamine neurons: The longer one is alive; the more dopamine neurons are going to die. The death of dopamine neurons underlies many of the symptoms of Parkinson's disease. Thus, if we all lived a really, really long time we would all get Parkinson's disease. Dopamine neurons are vulnerable to oxygen because they contain lots of iron. The iron inside these dopamine-containing neurons literally rusts, leading to the death of the neuron. Ironically, as a consequence of your addiction to food, you slowly kill off your dopamine neurons. The slow loss of these dopamine-containing neurons has another unfortunate consequence: age-related *anhedonia*, or the loss of interest and motivation to seek out pleasurable experiences. Research has shown that anhedonia increases with age. Your daily eating, breathing, and exercising contribute to this process. Obviously, this presents quite a conundrum if you are interested in remaining healthy since you must eat, you must breathe, and you must move in order to find something to eat; all so that you can grow old.

### Getting the balance right

In summary, it may seem paradoxical that exercise simultaneously protects the brain while also increasing the production of harmful ROS. The solution to the paradox is to understand that the amount and nature of the exercise is related to its benefits or harm for the brain and body. Moderate levels of exercise lead to metabolic mechanisms inside of your cells that normalize and neutralize the ROS levels and reduce oxidative stress, whereas intense exercise, or a single bout of extreme exercising by an inexperienced body, fails to induce these protective compensatory processes, leading to oxidative injury to the brain. Aerobic training increases the brain and body's own antioxidant abilities by upregulating the expression of critical antioxidant enzymes that are able to neutralize the oxidative damage caused by exercising. For example, ROS can induce the oxidation of body fats, making them quite toxic to other cells. Moderate levels of daily exercise increase the level of the enzyme glutathione transferase, which helps to protect your cells from the byproducts of fat oxidation. Moderate exercise also increases the expression of BDNF in the brain; this may make the neurons more resistant to oxidative insult. On the other hand, high-intensity exercise leads to ROS overproduction, which can cause mitochondrial dysfunction, reduced neurogenesis in the hippocampus, and neuronal death. Thus, the negative consequences of extreme exercising are only partly due to low levels of BDNF.

### The benefits of brain BDNF

BDNF is produced by neurons within the brain. It is thought to be involved in plasticity, neuronal survival, and the formation of new synapses, dendritic branching, and the modulation of both excitatory and inhibitory neural activity. It is active at all stages of development and aging. Studies on animals report an increase in brain BDNF levels when animals are exposed long term to environmental enrichment or physical activity. BDNF has been located in regions of the brain that control learning, memory, mood, eating, drinking, and body weight. Unfortunately, numerous attempts to study the role of BDNF in neurogenesis, neuroplasticity and development have yielded contrasting results. Overall, our current understanding of the role of BDNF in adult neurogenesis is highly contradictory. Thus, it remains difficult to define precisely the importance of exercise-related changes in brain BDNF levels.

Long-term moderate aerobic exercise is consistently correlated with elevated levels of BDNF in the hippocampus, as well as in structures related to the control of movement. However, changes in BDNF levels are not specific to exercising. Many recent studies suggest that BDNF signaling in the brain also mediates some of the beneficial effects of calorie-restricted diets. Thus, BDNF levels in brain will increase whether you exercise or simply eat much less food. Resistance exercise, such as weightlifting, also increases hippocampal BDNF levels. The elevation in BDNF levels can be quite long-lasting; for example, BDNF was still elevated in the hippocampus one month after the end of the exercise routine. Furthermore, the duration of the exercise does not always correlate with the change in BDNF levels. In addition, there does appear to be a threshold level of exercising that is required, since short-term running for only a few days produced no increase in BDNF levels. Thus, daily moderate exercise is required in order to induce the beneficial physiological changes in the hippocampus that support the processes of learning and memory. The benefits induced by BDNF may also depend on the presence of another protein called *nerve growth factor* (NGF), particularly as related to changes in neurogenesis and neuroplasticity. The level of NGF in the hippocampus was increased after eight weeks of treadmill running. It is highly likely that many different hormones and neurotransmitters in the brain are also altered by exercise; scientists have simply not yet discovered these changes. Thus, it is best not to place too much emphasis on the actions of BDNF alone in the brain. Part of the problem is that BDNF comes in so many different forms.

Understanding how BDNF acts in the brain is challenging because it is produced in so many different forms and each form appears to have slightly different functions. In the brain, proBDNF is produced first and is then converted into what is called the *mature* form of BDNF. The mature form is next converted into a molecule called *BDNF pro-peptide*. The BDNF pro-peptide plays a role in the specialized electrical process in the brain called *long-term depression* (this is not a form of mental disorder), which is an activity-dependent reduction in the efficacy of neuronal synapses. In a way, long-term depression helps with the consolidation of new memories by weakening older memories.

The level of BDNF pro-peptide in the cerebrospinal fluid is significantly lower in patients with major depressive disorder. The changes in BDNF pro-peptide may help explain sex differences in the incidence of major depressive disorders since male patients had significantly lower levels of BDNF pro-peptide in their cerebrospinal fluid than did the depressed female patients. Thus, the BDNF pro-peptide is a biologically important synaptic modulator that may also be associated with major depressive disorder in males.

Unfortunately, the precise function of these different versions of BDNF in other psychiatric disorders is still unknown. What *is* known is quite interesting. The levels of mature BDNF in the parietal cortex (located in the back half of the brain) from patients with major depressive disorder, schizophrenia, or bipolar disorder were significantly lower than in the brains of a control group, whereas the levels of BDNF pro-peptide in this area were significantly higher than controls. These preliminary findings suggest that abnormalities in the processing of BDNF from its gene into a fully mature protein represent a more important predictor of major psychiatric disorders than a simple measure of total BDNF levels. Too often in the past researchers have overemphasized baseline levels of only the mature protein. I believe that this has led to misleading conclusions about the role of BDNF in the brain. Thus, the ability of exercise to increase levels of BDNF in the brain may not impact mental health unless the processing of BDNF can be restored to normal. It is currently not known whether exercise is capable of altering how BDNF is processed into its mature form.

Exercise does increase the production of BDNF in the hippocampus. However, as I discussed earlier, there is no evidence that BDNF is originating in the periphery and then entering the brain. Fortunately, this does not matter since your brain can make its own BDNF. BDNF production in the hippocampus happens because the gene for BDNF, which is usually turned off, is allowed to be translated and transcribed into the protein precursor for BDNF. The regulation of the BDNF gene in the hippocampus depends on the modulation of specialized proteins, called *histones*, that typically prevent the reading of the DNA code. Think of histones as blankets that cover up sections of your DNA that contain specific genes. Exercise does not induce the expression of genes directly; apparently, exercise induces changes in where, and how much, your blanket of histones is pulled away to uncover specific genes so that they can be translated into proteins such as BDNF. Essentially, exercising leads to elevated brain BDNF levels by reducing the action of these inhibiting proteins on the BDNF gene—exercise inhibits the inhibitor.

Exercise might also influence the levels of BDNF in the hippocampus via another pathway. Endurance exercise induces the production of the ketone beta-hydroxybutyrate, one of the breakdown products of the metabolism of body fat for energy by your muscles. Exercise leads to the accumulation of beta-hydroxybutyrate in the hippocampus. This ketone inhibits the same class of histones mentioned earlier, leading to elevated levels of BDNF in the hippocampus as an indirect consequence of exercise. The changes in BDNF expression are only seen in the hippocampus and not in any other brain region. It is not known why the response to beta-hydroxybutyrate is localized only to the hippocampus. One possible explanation is that exercise selectively activates ketone transporters only along the section of the blood–brain barrier that surrounds the hippocampus. Consistent with this explanation is the discovery that exercise induces the expression of proteins only within the hippocampus that can selectively transport beta-hydroxybutyrate across the blood–brain barrier. The increased ketone transfer into the hippocampus occurs immediately after exercise, and the effect can last for up to 10 hours. What is particularly interesting is that the exercise-induced upregulation of a ketone transporter was also associated with increased levels of BDNF in the hippocampus. These same effects on hippocampal levels of BDNF were achieved when beta-hydroxybutyrate was injected directly into the brain.

What do these findings suggest with regard to how exercise influences brain function? Taken together, these results are consistent with the hypothesis that the benefits of aerobic exercise on increased BDNF levels, and the adult neurogenesis that it might induce, depend on the entry of beta-hydroxybutyrate into the hippocampus. It is especially important to appreciate that these beneficial changes in brain chemistry only occurred following aerobic exercising. Strength-enhancing exercises suppressed aerobic exercise–induced cognitive improvements and hippocampal neurogenesis; furthermore, greater intensities of the strength-enhancing exercises produce greater levels of suppression. Low- and high-intensity strength-enhancing exercises (e.g., weightlifting) will actually impair the aerobic exercise–induced increases in beta-hydroxybutyrate and BDNF, as well as the cognitive benefits and hippocampal neurogenesis! Please do not jump to the conclusion that weightlifters are less intelligent than runners due to the reduced level of neurogenesis within their hippocampus; they simply do not obtain the same degree of benefit for their brain from their sport.

At this point, it is safe to assume that exercise indirectly elevates BDNF in the brain via the actions of a variety of other hormones and proteins. Are these extra BDNF molecules in the brain responsible for the cognitive benefits associated with exercise? No one knows, yet. How BDNF acts to improve cognitive function remains a mystery. However, neuroscientists have discovered some potential mechanisms. One possible explanation is that BDNF increases the volume of a region of the hippocampus that is critical for learning and memory; this region is called the *dentate gyrus*. The increase in volume of the dentate gyrus is due to the growth of new blood vessels. The new blood vessels develop in response to the elevation in yet another hormone; this hormone that induces the growth of new blood vessels is called *vascular endothelial growth factor* (VEGF). VEGF is induced by exercise.

Following regular daily moderate exercise, the improvements in learning and memory abilities are most likely due to the augmentation of the blood flow to regions of the brain that are responsible for these abilities. Overall, studies using various animal models have demonstrated that the benefits of moderate aerobic exercise for the brain, such as reduced indicators of oxidative stress and increased levels of BDNF, are most likely due to an elaboration of blood vessels leading to regional enhancements in blood flow. The increased blood flow also reduces the level of inflammation due to ROS formation. In contrast, reduced blood flow to the brain associated with certain diseases, with normal aging, or following extreme endurance exercising is associated with poorer learning and memory abilities, impaired executive functioning, and mood changes. Researchers found that after several weeks of overtraining, athletes became more likely to choose immediate gratification over long-term rewards. The finding could explain why some elite athletes see their performance decline when they work out too much—a phenomenon known as *overtraining syndrome*. Overtraining syndrome is a form of burnout that leads to a decrease in performance that is often associated with feelings of intense fatigue. Inside the brain, extreme exercising resulted in significantly reduced activation of the cortex located in the left side of the frontal lobe called the *medial prefrontal cortex*. This brain region is thought to control your ability to resist the temptation of accepting a small immediate reward versus waiting for a bigger reward somewhat later. The cognitive control fatigue may be related to the increased release of pro-inflammatory cytokines, which are known to affect motivational processes.

How does inflammation alter how your brain functions? This is a question that I have spent the past 20 years trying to answer. If you imagine that your brain cells connect to each other with wires, inflammation causes the wires to disconnect. What is particularly interesting is that some regions of your brain are more vulnerable to the consequences of inflammation than are other regions. My research demonstrated that the hippocampus and some regions of the cortex are the most vulnerable to the negative consequences of the type of inflammation that is produced by extreme exercising.

### Calming the brain inferno with exercise

One potential long-term benefit of improved blood flow to the brain in response to exercise is a reduction in the level of inflammation. Poor perfusion of the brain, coupled with increased levels of ROS, leads to increased levels of pro-inflammatory proteins. In addition, physical inactivity is associated with increased viral and bacterial infections, leading to a systemic low-grade inflammation; regular exercise reduces the incidence of these infections. Overall, regular moderate levels of exercising enhance the body's immune function and anti-inflammatory processes. Thus, inflammation is not always harmful; what matters is the balance of its effects on the brain.

Your inflammatory response is a complex series of defense mechanisms that evolved to protect your body from infection and injury by inducing your body to release proteins, such as the cytokines I mentioned earlier, that initiate a cascade of biological processes designed to help you to defend from invading organisms and toxins, as well as to initiate the healing process. These proteins can have both beneficial and harmful effects; it depends on what is causing the inflammation to develop. One of the most common causes of body and brain inflammation in otherwise healthy people is obesity. Obese people have significantly elevated levels of cytokines in their blood. The inflammatory cytokines enter the brain easily. Once there, they induce shrinkage of brain regions (primarily gray matter [where the neurons live]) that are used in the process of learning new things and recalling memories. As I described earlier, cytokines are capable of literally disconnecting the physical links between neurons, called *synapses*, which are the basis of learning and thinking. This explains the increased levels of inflammatory proteins floating around inside the brains of obese people, directly impairing their learning and memory abilities.

### Liposuction versus exercise

What would happen if some fat cells were simply removed? Would the inflammation level decrease? Exercise can shrink fat cells, but liposuction can remove them from the body. Can liposuction make you smarter? A group of scientists recently investigated this novel question by conducting three very clever experiments on obese and normal-weight mice.

First, a group of obese mice were forced to exercise on a treadmill. Unlike the millions of Americans who own treadmills, these mice had no choice but to run. As expected, the daily treadmill exercising reduced belly fat, reduced the level of inflammation in their body, and significantly restructured how their brains function at the cellular level, leading

to improved memory. In a parallel study, the scientists surgi-
cally removed fat pads from a similar group of obese mice—in
other words, they underwent a standard liposuction proce-
dure. The results were identical to those produced by running
on the treadmill: Inflammation was reduced and the mice performed significantly better on tests of learning and memory.
These findings confirm many recent studies that have documented the ability of fat cells to impair brain function and
accelerate aging.

Then the scientists did something truly astonishing: They
transplanted fat pads into normal, healthy-weight mice. The
impact of the fat cells was immediately obvious. The mice
showed increased signs of brain and body inflammation, and
they developed deleterious changes in brain structure and
function that led to reduced memory performance. The inser-
tion of fat pads from another rat made these rats stupid.

Today, an overwhelming amount of scientific evidence
obtained across a wide spectrum of medical disciplines
strongly argues that obesity accelerates brain aging, impairs
overall cognitive function, and, ultimately, is responsible for
the numerous biological processes that kill us. This recent
study suggests that the simple removal of excess fat cells can
produce significant health benefits that equal or surpass the
benefits of exercise.

Fat cells full of fat globules are so toxic that their removal
provides health benefits! Liposuction is not an option for
everyone; however, the results of this area of research indicate that the biggest health risk to the brain is obesity. When
you are young, it is possible to be obese and healthy, but this
safe period does not last for long. Ultimately, your excess
body fat will make you sick, induce diabetes, accelerate arthritis, induce and then feed a variety of cancers, and then
kill you.

In this study, obesity was harmful to the brain function of
mice, but does it also impair the brain function of humans? In
order to investigate this important question one recent study

examined the relationships between academic performance, cognitive functioning, and body mass index (a ratio of body weight to height) among 2,519 young people. As predicted by numerous animal studies, the body mass index of humans was inversely correlated with general mental ability even after controlling for demographics, lifestyle factors, and lipid profiles. Thus, the greater the level of obesity, the greater the intellectual decline. Specifically, this study determined that obesity worsens performance in what are called *cognitive control tasks*. For example, having cognitive control means being able to override a bad initial impulse. For example, you are hungry and you see a piece of chocolate cake on your roommate's desk, but you don't eat it; it is not healthy, and it is not yours. Having cognitive control also means being able to ignore things around you that are not relevant to your current task. Finally, something everyone is familiar with today, having good cognitive control means that you can perform multiple tasks at the same time, such as writing a text message to someone while talking to someone else on the phone. If you cannot do any of these things as well as you would like, then you should consider losing some body fat and the brain inflammation that it produces and that underlies your loss of cognitive control.

And you should do it soon! The longer the inflammation is present, the more brain shrinkage occurs, and the more cognitively impaired you will become. This is why it is so important to avoid obesity in childhood. Being obese at midlife is also a strong predictor of dementia in later life. As expected, elderly obese people have more impaired learning and memory abilities than elderly thin people. Thus, inactivity and long-term consumption of high-calorie diets combine to injure the brain and impair both mood and cognitive function. Once again, the best advice is moderation in all things.

This is your new mantra:

*Shrink the fat before it shrinks your brain.*

## The hazards of thinking too fast

Neuroscientists have discovered that a healthy balance is also necessary for thinking, just as it is for exercising. The major culprits underlying our vulnerability to increased brain and muscle activity are oxidative stress and inflammation. Inflammation worsens the negative consequences of oxidative stress within the brain. Because the brain uses more energy than any other organ in your body, the mitochondria inside its billions of neurons produce a lot of ROS. Neurons face injury from the inside via increased levels of ROS and from the outside due to the presence of elevated levels of inflammatory proteins; the combined negative consequences of this environment place neurons at increased risk of oxidative damage and death. ROS production varies in parallel with brain activity: The more active your brain becomes, the more ROS it produces. Recent studies have identified the biological mechanisms that underlie the death of overly active neurons in the hippocampus. By now you might have noticed that the hippocampus is highly vulnerable to the consequences of exercising, breathing, eating, and thinking. Thus, it is not surprising that the hippocampus shows degenerative changes during the earliest stages of age-related brain disorders such as Alzheimer's disease. Memory loss due to hippocampal dysfunction is one of the most characteristic symptoms during the initial stage of the disease. Neuroscientists now understand why the hippocampus is so vulnerable to chronic neuroinflammation and oxidative stress. In addition, epidemiological studies have discovered an interesting correlation between overusing your hippocampus and becoming demented in old age. For example, the decline in cognitive function of patients with Alzheimer's disease is faster, after diagnosis, in patients who attended college versus patients who did not attend college. Think about the irony of this (but not too much, because doing so might kill some neurons!). When it comes to your muscles, the mantra is "use it or lose it," but when it comes to your brain, the new mantra is "use it too much and you will lose it." Thus, you should also take time every day to sit quietly and rest your mind in some form of contemplative silence, such as mindful meditation, to maintain good mental health and reduce ROS production. And try to get some sleep.

## Sleep and exercise

Disturbed sleep patterns represent a significant public health concern that may negatively affect your brain almost as much as long-term obesity. Indeed, poor sleep habits, such as staying up too late every night and not getting enough sleep, predispose us to obesity. When my students talk about their sleeping habits during class, most of those who exercise regularly claim that their sleep quality ranges from fairly good to good and will argue that the myths about the negative effects of exercising prior to bedtime do not apply to them. Does exercise help with insomnia? A meta-analysis of over 50 recent studies found that regular exercise, regardless of its intensity or aerobic/anaerobic classification, had a modest to moderate benefit on overall sleep quality, and a small benefit on total sleep time and sleep efficiency. In addition, the data obtained from many recent studies suggest that the modest sleep-enhancing effects of exercise may not be noticeable initially. Overall, exercise does benefit sleep quality, but the overall benefit may take time to evolve fully and may not be as substantial as many would hope.

Good sleep is an elusive commodity that becomes harder to achieve with age and is sensitive to the bias of the person who is attempting to get a good night's sleep. Your ability to rate how well you slept the previous night is greatly influenced by how easily you fell asleep and by how rested you felt on wakening. Sleep scientists have discovered that these are not always reliable or valid indicators of your overall sleep quality. This makes studying sleep quite challenging. Sleep scientists do not completely understand how physical activity affects your sleep quality or how long you sleep each night, although some specific genes that regulate your biological clocks might be involved.

## Muscular clocks

Obviously, good sleep contributes to good mental health. Although multiple brain regions are known to control the mechanisms and rhythms of sleep, the brain does not contain the only clock that sets the daily sleep–wake rhythms. Circadian clocks, and the genes that control them, have been located in nearly every cell of your body. Muscles have their own biological clock genes that are thought to be regulated by the body's other circadian clock genes. If true, this predicts that there are daily time windows during which exercise may exert its maximal beneficial effects on brain function. This relationship may work both ways: Exercising affects the biological mechanisms of circadian rhythm, and these rhythms affect muscle physiology and the potential balance between the beneficial and negative effects of exercising. So, what is the connection between your muscles and your biological rhythms?

Your muscles contain a clock gene; it is called *Bmal1*. Clock genes are components of the circadian clock comparable to the cogwheels of a mechanical watch. Bmal1 is actually attached to your DNA inside the nucleus of the cell. Bmal1 plays a critical role in many diverse biological functions, including the control of certain stages of sleep. The importance of Bmal1 was demonstrated by a study that investigated what happened to sleep quality when the Bmal1 gene was removed from skeletal muscle. The loss of the Bmal1 gene inside of muscles impaired the pattern of sleep cycles at night. Another study investigated the exact opposite question: What would happen if the Bmal1 gene was overexpressed in skeletal muscle? Animals with extra copies of the Bmal1 gene became resistant to the behavioral effects of sleep deprivation. Sleep deprivation is something that everyone has experienced many times; it leaves us tired and irritable, and we find it hard to concentrate on daily tasks that are usually quite easy after a good night's sleep. Apparently, this circadian clock gene in the muscles of your body influences how well you sleep at night and how well you can tolerate a bad night of sleep.

Exactly how the Bmal1 gene is able to influence the brain's control of sleep rhythm is not yet known; nevertheless, the results from animal studies are intriguing because they demonstrate that a gene in skeletal muscle can influence sleep processes that were once thought to be solely under the control of the brain. The implication is that an important muscle-to-brain communication pathway exists and might explain some of the benefits of daily moderate exercising on healthy sleep. In contrast, endurance athletes are well aware of the negative effects of excessive exercising on the quality of their sleep. Once again, finding the best balance between amount and duration of exercising matters to brain health.

The timing of the exercising and its effects on sleep may offer some insights into the physiological mechanisms underlying their relationship. For example, exercise heats up the body, and this can interfere with falling asleep. In contrast, exercising a few hours before trying to fall asleep allows the body time to rebound into a post-exercise drop in temperature that would promote falling asleep. Overheating may explain why falling asleep can be difficult after a major endurance competition or a long and strenuous workout, despite feeling thoroughly exhausted. There are many reasons for this form of insomnia. After a strenuous workout, the body tends to radiate heat due to the release of peripheral hormones. Due to the effects of ambient temperature, falling asleep in January is usually much easier to accomplish than falling asleep in August. At the beginning of a healthy sleep cycle your body needs to cool down. Research has shown that this cooling is a necessary component of the transition from waking to the initial critical stages of sleep called *slow-wave sleep*. Slow-wave sleep appears to be controlled by the clock muscle gene Bmal1. That is why good sleep habits require that you participate in some form of moderate exercise during the day and then make certain that the bedroom is as cool, quiet, and dark as possible.

Falling asleep is also much harder to achieve when the blood levels of the stress hormone cortisol are elevated. During the few days following excessive exercising routines, elevated levels of cortisol, combined with poor sleep quality, can induce mental fatigue, slow reflexes, impair decision making, and induce headaches and dizziness, as well as significant mood changes, such as depression, anxiety, and irritability. In addition, endurance exercising induces the release of adrenaline from the adrenal glands, producing feelings of enhanced arousal for many hours. In contrast, mild to moderate levels of exercising do not increase the time it takes to fall asleep or impair sleep quality. Poor sleep quality can induce feelings of low energy, fatigue, and depression.

# 12

# EXERCISE FOR THE DEPRESSED BRAIN

You may be reading this section before any of the others because you, or someone you care about, suffers with symptoms of depression. Depression has been referred to as the "common cold of mental illness" because virtually everyone gets it at some time in their lives. Because of its high incidence in the general population, and because the popular literature claims that exercise can effectively reduce the symptoms of depression, I am going to discuss this topic in depth.

How do we treat depression today? Most often depressed patients are given drugs that enhance the function of serotonin, norepinephrine, or dopamine. When these neurotransmitters are released by a neuron that binds to specific protein receptors, they usually induce a change in function of the next neuron and then quickly disconnect from the protein receptor. What happens next is what matters here: The neurotransmitter detaches from the next neuron and floats around inside the space between neurons. The neuron that released the neurotransmitter then vacuums up about 80% to 90% of the serotonin or norepinephrine or dopamine molecules, thus effectively inactivating them. The remaining neurotransmitter molecules that are not vacuumed up are usually destroyed by enzymes; the byproducts of this minor process can be found in the cerebrospinal fluid and urine. The vast majority of antidepressant drugs do one thing: They selectively block this reuptake (vacuuming) process. That is why they are called *selective reuptake inhibitors*. A drug that is a selective serotonin reuptake inhibitor is called an SSRI.

Please consider what happens immediately after you swallow your SSRI. Within about one hour, it has entered your brain and immediately starts blocking reuptake of serotonin. Do you feel better? No. So, how long do you have to wait to feel better? Two hours, ten hours, twenty-four hours? Nope. You must keep taking these drugs for many weeks and then, maybe, if this is the correct drug therapy for you, you begin to feel better. Thus, the blockade of reuptake never coincides with a reduction in the symptoms of depression.

Are these selective reuptake inhibitors effective? One recent publication summarized what most patients experience: Currently available treatments for depression are modestly effective, while treatment resistance and recurrence of symptoms remain significant problems. The sad truth is that our currently available medications are no more effective than those introduced over 50 years ago. Complicating the answer to the question of whether these drugs are effective is the fact that 23% of all depressed adults will experience spontaneous remission without treatment within about three months. Indeed, about 32% will spontaneously recover in six months and 53% will spontaneously recover from their depression in a year. Thus, if you can tolerate the symptoms for one full year, you have a 50:50 chance of getting better by doing nothing. For most people, that is not an acceptable solution, however. Just waiting places the patient at risk of committing suicide before they recover. Also, numerous studies have shown that people whose depressive symptoms are severe are much less likely to get better without treatment. Unfortunately, about one-third of all depressed patients never respond to any therapy. The most common residual symptoms for these unlucky people are insomnia, fatigue, painful physical complaints, problems concentrating, and lack of interest in pleasurable activities. Medical science clearly needs to discover better therapies for this disorder that affects so many people around the world.

Does exercise reduce the symptoms of depression? The answer to this question parallels the statements we just made. Exercise is modestly effective for a small percentage of people; however, for those with more severe symptoms, exercise alone is much less likely to be effective. In addition, it is difficult to isolate the benefits of regular exercise from any other daily activity because a significant percentage of patients will have their symptoms resolve by doing nothing at all.

### The role of BDNF

How might exercise reduce the symptoms of depression? Let's examine the popular explanation that the benefits of exercise on depression depend on increasing the level of BDNF in the hippocampus. As I discussed earlier, BDNF plays a critical role in the growth, survival, and differentiation of the developing brain. In support of this explanation are reports of a correlation between decreased production of BDNF in the hippocampus and the symptoms of depression. A similar statistical correlation has also been described between increased levels of BDNF in the brain and serum, regular exercising, and a reduction in the symptoms of depression. While these correlations are intriguing, no study has yet described a mechanistic link between depression and reduced BDNF production within the hippocampus. Many years ago, a correlation was discovered that linked a mutation of the BDNF gene with numerous mental disorders, including bipolar disorder and panic disorder; however, the decline in function associated with this mutation did not induce depression in adult humans. As happens so often in science, the earliest studies reported correlations between BDNF dysfunction and depression that cannot be replicated in subsequent studies.

Probably one of the most interesting recent discoveries is that treatment with antidepressant drugs, such as the popular SSRIs or selective norepinephrine reuptake inhibitors (SNRIs), correlates with an increase in the concentration of BDNF in the hippocampus. At about the same time that this discovery was published there were some preliminary reports that increased exercise was correlated with increased BDNF levels in the blood and brain. These initial reports seemed so conclusive that the United Kingdom's National Institute for Health and Clinical Excellence published a guide that recommended treating mild clinical depression with exercise rather than SSRI antidepressants. This recommendation might have been ill advised because it was not adequately supported by all of the available data. Furthermore, relying on the potential benefits of exercise alone to treat depression could potentially be dangerous for suicidal patients, who might obtain more protection by depending on standard pharmaceutical treatments, with or without regular exercising. Yes, exercise did improve the mood of some mildly depressed patients; however, this preliminary finding was not adequate proof to defend their imprudent decision to recommend that exercise is more effective than standard antidepressant therapy.

In the United States, the officials at the National Institute of Mental Health (NIMH) have been more skeptical due to multiple concerns about potential experimenter bias and problems with the internal and external validity of the initial studies that made these claims about the benefits of exercise on depression. Essentially, the NIMH did not trust the results of these studies.

In summary, reduced BDNF is not a specific or reliable diagnostic marker for depression. Currently, no one is sure whether low BDNF levels lead to depression or whether depression causes BDNF levels in the brain to decrease. If exercise is responsible for a reduction in the symptoms of depression, there is no solid causal evidence that this is due to increased brain levels of BDNF. The current situation is a version of that old causality dilemma of the chicken and the egg; no one knows which came first.

## Why do our antidepressant drugs fail us?

One of the reasons that the NIMH did not trust the original studies was that these authors made the typical assumptions about the underlying causes of depression; then, they explained their results in terms of these underlying causes. This is typical circular logic, and it usually goes something like this:

1. Antidepressant drugs and exercise improve mood by enhancing the function of serotonin.
2. Thus, depression depends on a dysfunction of serotonin that can be corrected by either exercise or antidepressant drugs.

I do not accept this logic as a valid explanation of the mechanisms underlying depression. This is why.

One of the strongest arguments to support the role of serotonin in depression comes from tryptophan depletion studies that were initially performed on animals almost 25 years ago. Essentially, laboratory animals or humans are fed a diet high in large neutral amino acids; this diet effectively reduces the production of serotonin in the brain, leading to symptoms of depression, sometimes within only a few hours. These results were used to defend the decision to administer SSRIs and tryptophan to people with depression. The problem with the conclusions drawn from the tryptophan depletion studies is that they explain very little about how depression develops in the brain of so many normal people. Overall, the tryptophan depletion model is not valid and makes inaccurate predictions about the effectiveness of drugs that enhance serotonin function. It does not explain why depression is incredibly common. Should we conclude that the millions of people who have been diagnosed with depression have inadvertently restricted their tryptophan intake? This would actually be a very difficult thing to do; the amino acid tryptophan is contained in many different foods.

The tryptophan depletion studies demonstrate an important feature of brain chemistry that actually undermines its validity. It demonstrates that it is possible to quickly slow the production of serotonin, and observe the behavioral consequences, by restricting the brain's access to its precursors in the diet. However, and I cannot emphasize this critical aspect of brain function too much, simply because the abrupt depletion of tryptophan in the brain *can* reproduce the symptoms of clinical depression *is not proof* that this is actually what occurs in the brains of depressed humans. The tryptophan depletion studies predict that SSRIs should be very effective for most people; the clinical evidence certainly does not support that prediction.

The activity of serotonin neurons in the brain is often determined by examining for changes in the level of serotonin's main metabolite (a chemical that your brain converts serotonin into so that it can be excreted by the body) in samples of cerebrospinal fluid. Depressed people have too few metabolites in their cerebrospinal fluid. These findings appear, at first glance, to be compelling support for the role of serotonin in depression. After all, the popular literature is full of claims that serotonin controls how you feel. The claimed relationship is clear and simple: Too little serotonin equals depression, while lots of serotonin equals happiness.

This rationale for using SSRIs has been lectured to physicians for the past couple decades. It is logically appealing and simple to understand. However, the actual chemistry of depression is much more complicated. Today, if you tell your doctor that you are depressed, he or she is going to write you a prescription for a drug that enhances the function of serotonin in your brain. Taken together with the results of studies that have examined serotonin metabolites after exercise, it seems obvious that the elevation in the level of serotonin metabolites in the cerebrospinal fluid can only mean one thing: Exercise and SSRIs improved your mood by enhancing serotonin function in the brain. Once again, correlation never proves cause and effect.

What these original studies did not consider is that the changes in serotonin metabolites in cerebrospinal fluid are more likely due to other factors related to mixing of cerebrospinal fluid metabolites from regions of the brain that are not related to mood control. For example, serotonin plays a role in the function of the motor cortex, a brain region having nothing to do with your mood. When your motor cortex is active, for example when you are exercising, increased levels of serotonin metabolites will collect in the cerebrospinal fluid. However, this increase only demonstrates a well-known link between the control of movement and the release of serotonin from nerve terminals in the motor cortex. Unfortunately, this correlation tells us little about any possible relationship between exercise-induced serotonin metabolites in the cerebrospinal fluid and changes in your mood.

Many studies have reported a correlation between serotonin metabolites in the urine with exercising. Unfortunately, monitoring serotonin metabolites in the urine is also unreliable because 99% of the body's serotonin does not originate in the brain. Most of the serotonin in the body is contained within your intestines and blood platelets. Thus, fluctuations in serotonin metabolites in the urine do not accurately represent parallel changes in brain serotonin function. This makes correlations between serotonin metabolites in the brain and exercise tenuous, at best.

Exercising has also been correlated with increased levels of plasma tryptophan. Tryptophan is a precursor in the production of serotonin. This finding also seems exciting at first glance and has led many authors to conclude that a rise in tryptophan availability in your blood following exercise leads to an increase in serotonin synthesis in your brain. This is simply not true; the brain does not function in such a simplistic manner. The brain is only able to utilize tryptophan in the blood in order to synthesize serotonin if the tryptophan is able to cross the blood–brain barrier; tryptophan cannot cross the blood–brain barrier without assistance. In addition, the transit of tryptophan across the blood–brain barrier is impeded when the blood contains other amino acids that compete with its uptake. This happens when you consume any protein; tryptophan makes up only a small percentage of the amino acids in most meats. In fact, most of the tryptophan that appears in the blood after a typical meal or following exercise will never enter the brain; rather, it will be utilized by one of the 10 million or more serotonin neurons in your gut, or by one of the trillions of platelets in your blood, or by any one of your muscles. In fact, far less than 1% of the tryptophan that you consume will ever reach your brain. Overall, the current correlational evidence linking exercise to enhanced serotonin function is not compelling; it becomes far less compelling once you understand how serotonin neurons in the brain actually function.

Once the tryptophan does enter the brain, it can become absorbed by serotonin neurons; so far, so good. Of course, it might also be absorbed by one of the many other cells inside the brain. The popular literature often gets the next steps in the process performed by serotonin neurons very wrong. Many authors claim that having more tryptophan inside of serotonin neurons will lead to improved serotonergic activity and elevated mood. Once again, it does not work that way. Merely because the serotonin neurons *make* more serotonin does not mean that they will actually *release* more serotonin. That, in a nutshell, is the fatal flaw in the logic of the popular literature. Serotonin must be released by your neurons in order for it to influence brain function. Simply producing more serotonin, but not releasing it on to the next neuron, will not influence how you feel.

Apparently, serotonin neurons produce lots of serotonin that they never release, and your serotonin neurons quickly destroy any extra unnecessary or unneeded serotonin that they produce. Animal studies have demonstrated that movement alone, even running in a circular maze, does not increase the firing rate of serotonin neurons. Only two stimuli have been shown to reliably increase the firing rate of your serotonin neurons; the two things that your brain evolved to care the most about: food and sex. Not exercise! Therefore, having more tryptophan in the blood and finding more serotonin metabolites in the brain and cerebrospinal fluid is not proof that the brain's serotonin neurons, particularly those involved in mood control, are responsible for influencing your mood following exercise. Never forget: Correlation does not prove causation. Fortunately, recent research may have resolved this confusion by uncovering a more plausible mechanism that links an elevation in mood and changes in BDNF levels with exercise and antidepressant drug treatment. I need to return your focus back to the effects of inflammation in the brain.

### Inflammation reduces BDNF

Exercise and antidepressant drugs reduce brain inflammation, particularly within the hippocampus. The reduction in brain inflammation leads to increased levels of BDNF in the hippocampus, and these changes correlate with less depression. The main benefit of antidepressants and exercise is now believed to arise from their shared ability to reduce brain inflammation. Consistent with this new insight into the role of brain inflammation in the cause of depression, a new generation of antidepressant drugs have been invented that are effective because they reduce inflammation. What is most interesting is that these novel antidepressant drugs do not alter serotonin function at all! Thus, scientists now think that the antidepressant action of the currently available prescription medications, such as the very popular SSRIs, might actually be primarily due to their ability to reduce the inflammatory environment within the brain (although see the reference at the end of the book by Lee et al., 2018). The antidepressant action of exercise is also most likely due to its ability to reduce inflammation within the brain and body. Reducing inflammation in the hippocampus allows BDNF levels to increase. As mentioned earlier, increased BDNF induces the development of new blood vessels via the actions of VEGF, thus increasing hippocampal blood flow; together, these processes induce neurogenesis in the hippocampus and improve cognitive function. These data are consistent with the hypothesis that depression is most likely a consequence of brain inflammation. However, although inflammation offers a valid explanation for the effects of exercise on brain BDNF levels, the complete story still requires some attention be given to the role of tryptophan in your brain.

### Inflammation and tryptophan

The beneficial effect of physical activity on depression may depend on a metabolic link between the metabolism of the amino acid tryptophan within muscles and within your brain. The link begins with how your brain and body metabolizes the tryptophan that is so common in your diet. Ordinarily, tryptophan is metabolized into kynurenine within your muscles. Kynurenine is quite toxic. At first glance, this does not seem helpful. Fortunately, kynurenine is quickly converted into kynurenic acid by an enzyme called *KAT* that is found in muscle. KAT is the hero in this story. Fortunately, exercise activates KAT, shifting metabolism toward the conversion of the toxic kynurenine into the safe kynurenic acid. Once formed in the muscles following the metabolism of tryptophan, kynurenic acid is unable to cross the blood–brain barrier. Thus, exercise reduces the level of toxic kynurenine in the blood as well as limiting the amount of kynurenine that can enter the brain.

However, your brain is capable of making its own kynurenine from tryptophan. Once again, KAT comes to the rescue and the kynurenine is metabolized into the less harmful kynurenic acid. Unfortunately, in the presence of brain inflammation kynurenine production is increased beyond the ability of your KAT to protect you. The failure of the conversion of kynurenine to kynurenic acid increases your risk of depression. This failure of appropriate tryptophan metabolism may explain why inflammation plays such a causal role in the development of depression.

Although there may be multiple neurobiological pathways through which inflammation might affect mood, activation of the kynurenine pathway is sufficient. Once inside the brain, kynurenine is converted into neurotoxic molecules that can induce depression. Insight into this mechanism was gained when scientists realized that depression frequently appears against the background of sickness when the inflammatory response is intense and long-lasting. The transition from simply being sick to feelings of depression is mediated by activation of the kynurenine metabolism pathway that leads to the formation of the neurotoxic kynurenine metabolites.

### Depression and inflammation

One particular inflammatory protein, interleukin-6, is thought to underlie the development of depression; when present at elevated levels, interleukin-6 predicts a poor prognosis. During depressive episodes, as interleukin-6 levels become higher, so does the severity of the depression. Interleukin-6 directly and negatively affects brain functioning and impairs the production of many different neurotransmitters, including, but not limited to, norepinephrine, serotonin, and dopamine, leading to a deterioration of cognition, most frequently occurring as a slower information-processing rate, weakening of working memory, and impaired concentration. These are all typical symptoms of clinical depression.

In summary, as neuroscientists learn more about the cause of depression, they find less and less evidence that serotonin plays either a unique or even critical role in the disease. It is only when serotonin levels in the brain are artificially reduced by drugs or tryptophan depletion that serotonin function is related to depression. These models of depression do not provide a valid reproduction of the biological mechanisms that lead to depression, which is why the insights gleaned from these models have not led to the development of effective antidepressants. Neuroscientists now understand that the most basic principle about drug design is this: simply because a drug provides some benefit by enhancing serotonin is not proof that the underlying disease symptoms are related to serotonin dysfunction. The development of an effective treatment for depression will require that drug designers give up on these outdated ways of thinking about brain function.

## Exercise and depression

The effects of exercise on depressed individuals have been investigated extensively. Unfortunately, the conclusions drawn from these studies in the popular press or social media have gone beyond what the actual data can defend. Past studies have numerous fatal flaws that undermine my confidence in their results. Previous studies have varied wildly in size, type of control group, methodological rigor, length of follow-up, and even the type of exercise modality. Randomized trials of exercise have generally ranged in length from six weeks to four months and typically emphasized aerobic exercise, although some studies on resistance training also have been conducted. The many differences in study design have contributed to the current level of confusion and misunderstandings about the benefits of exercise on depression.

Although many trials have been conducted on adults with major depressive disorder, only a few used high-quality methodologies in which the treatment allocation was concealed. Many of the other studies failed to comply with the standard intention-to-treat analyses. This means that the final analysis included every subject who was assigned to a randomized treatment. This type of analysis ignores noncompliance (the subjects exercised too much or not at all), protocol deviations (they performed the wrong exercise), and withdrawal from the study, or any of a number of potential things that might have happened to the subjects after they had been assigned to their study groups. The general problem is it is very difficult to convince human subjects to exercise consistently without introducing their own creative changes that often undermine the integrity of the study. The results of many of these studies were made difficult to analyze because they failed to include a control group in the design. Sometimes the most important variable, the degree of depression experienced by the subjects, was assessed by someone who was not blinded to treatment. Thus, the studies were vulnerable to experimenter bias.

For all of these reasons, the initial enthusiasm and some outrageous claims about the effectiveness of exercise in combating depression have not been supported by newer studies. Recent well-controlled studies have found only a "modest to moderate" antidepressant effect due to daily long-term exercising. In studies that used a standard head-to-head comparison, exercise was no better or worse than standard cognitive-behavioral therapy for treating depression—that is, just talking to someone was as effective as exercising. When researchers compared the effectiveness of exercise on adults who were given a standard antidepressant drug therapy, they found no significant benefit of one approach over another. This surprising discovery actually says a lot more about the overall lack of effectiveness of most drug therapies for the majority of depressed patients than it does about the potential benefits of exercise.

One of the largest and best-designed meta-analyses of this question was published in 2019. It collected data from over 4,000 publications and analyzed the best ones for evidence that regular exercise may reduce depressive symptoms in patients who had breast cancer, prostate cancer, mixed cancers, cardiovascular disease, coronary heart disease, hemodialysis, fibromyalgia syndrome, acute leukemia, other hematological malignancies, heart failure, HIV infection, multiple sclerosis, mixed neurological disorders, Parkinson's disease, stroke, ankylosing spondylitis, traumatic brain injury, or lupus. The results of this impressive study were consistent with those of other studies: Exercise is only modestly effective in reducing depression. Thus, it is worth noting the following.

### The benefits of exercise are real but small

Overall, the benefits of physical exercise for the brain are subtle but real; people who exercise reported 43% fewer days of poor mental health each month as compared to people who did not exercise and who were matched for physical and sociodemographic characteristics. What is meant by "poor mental health" needs to be defined clearly in order to appreciate these results. Poor mental health includes a very broad range of self-diagnosed conditions that include, but are not limited to, confused thinking, sadness, irritability, excessive fears, worries, anxieties, social withdrawal, or significant changes in eating or sleeping habits.

The greatest reduction in symptoms was associated with participation in team sports, cycling, and most aerobic and gym activities that lasted about 45 minutes and occurred three to five times per week. The overall value of these studies is somewhat weakened by the fact that they did not distinguish between the potential mental health benefits provided by interacting (such as talking) with others, which can be significant in these sports, versus the benefits due to physical exercise. However, these studies do suggest that if you are feeling down, go out and play with someone.

Many of these studies were often challenging to conduct and interpret because depressed people, particularly adults, find it difficult to begin or maintain an exercise routine. Depressed people may also lack the motivation to exercise due to symptoms related to being depressed, such as fatigue, low self-esteem, and loss of interest in participating in pleasurable activities. In addition, some antidepressant therapies induce significant weight gain, which may undermine the patient's enthusiasm for exercise.

Causality also needs to be considered when examining the benefits of exercise on adults. Causality requires an answer to the question: Is there a real relationship between cause and effect? For example, are active people less likely to be depressed because they are active or because they are simply not depressed? It is difficult to determine which condition came first. Taken together, these recent studies clearly demonstrate that exercise is not better than drugs but should be considered as an important adjunct therapy to any treatment plan for depressed adult patients. Do these same conclusions about exercise in depressed adults also apply to depressed children?

# 13

# EXERCISE FOR THE
# CHILD'S BRAIN

Children also suffer from depression. About 7 million children in the United States have been diagnosed with anxiety or depression. This number likely underestimates the actual number since many families may lack the resources to obtain a medical diagnosis. Depression in children also frequently goes unnoticed and thus untreated.

Does exercise help? An analysis of a series of randomized studies of children in late childhood and early adolescence discovered that the benefits of physical activity were small and not long-lasting. A recent larger analysis that included 3,500 children ages six to 10 years asked a quite interesting question related to depression and exercise in children: If exercise has a significant antidepressant effect, would being sedentary lead to depressive symptoms in children? This large investigation of thousands of children found no relationship between inactivity and the risk of becoming depressed. Children who do not exercise are not at greater risk of becoming depressed, as long as they do not become obese. Furthermore, unlike the situation for adults, being depressed as a child did not influence the likelihood of being physically active or inactive. In summary, exercise is not an effective treatment for children suffering from depression; however, physical activity may be modestly beneficial for children who suffer with only minor symptoms of depression.

The benefits of exercise for both children and adults convey some of the same complications associated with drug therapies for depression. For example, the benefits of exercise are variable in effectiveness and duration. For some people, exercise helps resolve some of their milder depressive symptoms, while for others it provides minimal benefits. The same is true for medications; milder symptoms tend to respond much better to drug therapies or placebos than do more severe depressive symptoms. In addition, also consistent with the benefits of drug therapies, there is a baseline amount of exercise that is required to achieve any benefits on mood. Recent studies have found no significant difference in treatment response between children who exercised three days per week and those who exercised five days per week. Furthermore, there appears to be a threshold level that can be achieved; more exercising does not translate into less depression. Today, more and more young children lead sedentary lives, particularly after age seven. The connection between inactivity and becoming overweight is obvious, yet there is comparatively little causal evidence about the effects of physical inactivity on brain health during childhood. (It is important to point out that the absence of evidence is not the same as having evidence of the absence of any effect.)

One recent study investigated the effect of exercise on math achievement, executive function, fitness, and body mass index in 505 children. The authors predicted that physical activity should improve fitness, reduce body mass index, and improve math performance. After nine months the researchers reported no significant effects on executive functions, fitness, or body mass index; only math performance improved. The lack of benefit for executive functions might be explained by the lack of effect of exercising on overall fitness or body mass index.

A recent, and much larger, study examined the effects of a seven-month curriculum of physical activity on executive functions in more than 1,100 10-year-olds across 28 different schools. Their results were compared with children in the same grade at 29 other schools who did not participate in any prescribed physical activity. The physical activity curriculum consisted of three parts: physical activity education lessons in order to demonstrate how to exercise safely and effectively, daily physical activity breaks during school periods, and assigned physical activity homework that each student was expected to perform daily. In total, these children had an additional 165 minutes of physical activity each day. Seven months of substantially increased physical activity in these young children produced no significant benefit on any measure of cognitive function. Although the small effects on cognitive flexibility and motor skills were not statistically significant, the authors concluded, in spite of their own results (sometimes scientists simply refuse to believe their own evidence because it does not fit into their preconceived assumptions), that engaging in a minimal amount of physical activity may assist brain health during childhood. When you read popular assessments about any health intervention, always watch out for these bias traps; don't allow the writer to lead you into one.

Another recent study using rats compared the impact of voluntary exercise that was initiated during either adolescence or early adulthood on cognitive performance and BDNF levels in the blood. The results, once again, argue against a direct link between BDNF in the blood and cognitive performance. Adult-initiated exercise enhanced cognitive performance on a few different tasks; in contrast, exercise that began in adolescence did not improve performance in these tasks. What was most surprising in this study was that adolescent-initiated exercise increased the expression of an array of genes in the hippocampus that are related to learning, including BDNF; in contrast, adult-initiated exercise did not affect these genes. These are brain indicators of the benefits of exercise that, obviously, could not have been obtained using human subjects and do raise the very important question: Are we missing something important in our studies of humans? The results using animals suggest that exercise initiated during adolescence affects brain BDNF levels without actually improving cognitive function.

A recent study compared the influence of aerobic exercise on brain cortical activity and cognitive performance in schoolchildren (8 to 10 years old). Cortical activity was reduced after an exercise class and no effect on cognitive performance was observed. Overall, studies of the benefits of exercise in humans, young or old, rarely produce statistically significant improvements on valid measures of cognitive performance. However, exercise does modify some anatomical measures of brain function in young children.

### Exercise and changes in white matter

Recently, scientists have investigated whether exercise induces detectable changes in the anatomy of specific brain structures. They examined the effects of physical activity on the microstructure of white matter tracts in 143 children between the ages of seven and nine years. "White matter" refers to areas of the brain that are mainly made up of nerve fibers with their myelin sheaths. The myelin sheath gives white matter its pale color and acts as an electrical insulator around the axon. Axons with their myelin sheaths connect brain regions to each other and carry nerve impulses between neurons. Most regions of white matter contain millions of axons. During development, as the brain matures, the amount of white matter increases, leading to the formation of functional connections between different brain regions. In addition, as the thickness of the myelin sheath grows, the efficiency of communication between brain regions also improves. Monitoring changes in the structure of the white matter provides insight into maturation of the brain.

Various environmental factors can negatively or positively influence the maturation of white matter. For example, babies who are neglected by their mothers during early development have reduced amounts of white matter. In the current study, young children who participated in the regular physical activity program showed increased white matter, suggesting more tightly bundled and structurally compact fibers, in a structure called the *corpus callosum*. The corpus callosum is a large fiber bundle of myelinated axons that connects the left and right hemispheres of your brain. This structure allows the two halves of your brain to integrate cognitive, motor, and sensory information. These findings predict that exercising while the brain is still developing and maturing may lead to changes that allow the two hemispheres, each with its own special abilities, to communicate more efficiently with each other. The beneficial effects of exercise on myelination, and the potential long-term consequences on cognitive abilities as these children mature, need to be investigated further.

Given these interesting results on white matter connectivity changes due to exercise, recent studies have investigated the benefits of exercise in adolescent animals exposed to brain radiation therapy. Therapeutic radiation often causes widespread changes in white matter that have significant negative consequences for cognitive function. For children with cancer, survival after brain radiation therapy, or chemotherapy as well, comes with a price: These kids have a higher risk of developing a wide range of neurological and cognitive disabilities. Animal studies have demonstrated that exercise can partially restore the number of myelinated brain fibers damaged by the radiation therapy and improve information processing speed. Hopefully, these same benefits can be reproduced in children following radiation therapy.

### Exercising parents

The benefits of having specific exercise-induced molecules floating around inside your brain and body can occasionally be passed on to the next generation. A study published in 2019 confirmed that spending time on a treadmill, just once or for up to 20 sessions, alters the expression of specific genes that can influence the expression of several other genes, including those related to brain function. The regulation of genes depends on the modulation of specialized proteins that either prevent or allow the reading of the DNA code in your cells. Exercise does not alter the actual DNA code; rather, it alters how different genes are activated. These types of changes to the DNA are called *epigenetic changes*.

Physical exercise by the father acts as an epigenetic modulator for his son. Geneticists have long known that DNA methylation patterns in paternal sperm before fertilization have the potential to alter epigenetic programming in future offspring. Performing an aerobic exercise protocol two times per week for three months (which is just about as long as the human male spermatogenesis cycle) reduced the DNA methylation levels in the sperm of healthy young adult men. Exercise causes significant changes in the methylation of genes related to neurological disorders such as autism, schizophrenia, Parkinson's disease, and Alzheimer's disease. The precise nature of these changes (whether they are beneficial or not) remains to be determined. It seems clear that fathers who exercise regularly modulate the DNA of their future generations. Male mice who exercised for 22 days prior to impregnation improved the spatial learning ability of their offspring. The benefits of the paternal exercise occurred in spite of a lack of changes in hippocampal neurogenesis and BDNF levels in the brains of offspring.

Another study using mice found that the BDNF levels in offspring of males who exercised was higher; however, this elevation was transitory and lasted for only a few days after birth. Thus, the physical activity of the father prior to conception modulates the epigenetic machinery in the brains of his offspring; this effect is more often observed in male offspring. The physical activity of mothers has been examined in animal models but the results are unclear due to one major confounding factor—the mothers are far more intimately involved in the early phases of infant care and development. The influence of this caretaking overshadows evidence correlating the potential benefits of maternal exercise on her offspring.

# 14

# EXERCISE FOR THE CHEMO BRAIN

Chemotherapy can produce a broad range of negative consequences. *Chemo brain* is a common term used by cancer survivors to describe a series of cognitive impairments, including confusion, difficulty with learning new skills or multitasking, and feelings of mental fogginess. The condition may be a consequence of the cancer's actions within the body or may result from the therapies used to treat the cancer and its symptoms. Chemo brain occurs more often following administration of higher doses of chemotherapy and is also more likely to develop if the brain is exposed to radiation during the therapy. During the recovery phase after treatment, patients often have low blood counts that might reduce the availability of oxygen to the brain. Many patients also report significant sleep problems that are often associated with impaired daytime cognitive function, irritability, and memory loss. Widespread inflammation that directly and negatively affects brain function also contributes to the symptoms of chemo brain.

Fatigue is a common symptom of chemo brain and may contribute significantly to the cognitive impairment. Fatigue is also a common symptom of brain inflammation associated with other diseases. Thus, it is not surprising that increased levels of pro-inflammatory proteins have been reported in the blood of patients with chemo brain. The degree of fatigue is often correlated with the severity of depressive symptoms experienced by cancer therapy patients; these symptoms are likely a consequence of increased serum levels of the cytokine interleukin-6.

Executive functions are often impaired as part of chemo brain. Executive functions are mental processes mostly happening in the cortex that control and regulate complex behaviors and actions. One of these executive functions, called *working memory*, that is lost is the ability to temporarily store, decipher, and apply newly acquired information. These many different symptoms explain why breast cancer survivors often report difficulties associated with returning to work, concentrating on new tasks, paying attention, multitasking, or fulfilling their regular roles in society. Fortunately, for most cancer survivors chemo brain tends to be mild and to dissipate over time.

Does exercise help? Exercise may mitigate the executive function and working memory problems associated with chemo brain. Physically active breast cancer survivors perform better on working memory, executive function, and attention tasks. Exercising is a well-known treatment for cancer-related fatigue, possibly via its ability to protect against body and brain inflammation. One recent study concluded that the benefits of exercise on executive function and memory were due to the reduction in fatigue. Unfortunately, this study failed to examine for changes in inflammatory markers in the body and brain. However, the study's finding is consistent with the hypothesis that fatigue and the consequences of inflammation are the mechanisms linking cancer treatment and chemo brain.

In spite of the recognized benefits of exercising, epidemiological evidence suggests that most breast cancer survivors participate in very little exercise each day. Many survivors report that pain is a common problem, especially in the first few years after treatment, and that the pain makes it difficult to exercise regularly. These breast cancer survivors are not alone in terms of the need to deal with daily energy-sapping pain. Indeed, pain, either modest or severe, is the most common neurological condition experienced by everyone, almost daily.

# 15

# EXERCISE AND PAIN

Long before our ancestors discovered the benefits of hot water decoctions of willow bark that led to the invention of aspirin, our bodies evolved the ability to diminish our experience of pain by releasing painkilling proteins into the blood. In the mid-1970s, research confirmed that our bodies and brains produce endogenous morphine-like peptides that are called *endorphins*. These peptides control our experience of pain by stopping the flow of pain signals into our brain. Some endorphins also act within the brain to modify our emotional response to the pain—you might feel the pain, but you just don't care that much about it. I have heard this same response from many of my student athletes.

Pain signals travel along neurons into the spinal cord prior to traveling up into the brain itself, where the pain is "experienced" by dedicated pain-processing regions. Within the spinal cord, the pain can be stopped, or greatly diminished, by the action of endorphins that are released locally near where the pain signals arrive. In addition, these endorphins are released into the blood, where they spread around your entire body to reduce the sensation of pain. The effect is similar to taking morphine. Aerobic and anaerobic training produce an increase in blood levels of endorphins in response to the intensity of exercise. Think about what this discovery tells you about intense exercising: It produces so much tissue damage that your body is compelled to release pain-relieving chemicals. As exercise intensity and duration increase, so do the blood levels of pain-relieving endorphins; more exercising-induced pain requires that more pain-relieving chemicals be produced and released. The endorphin increase depends on your health status (whether you are already in pain prior to exercising) and the type of exercise. For example, performing only one session of anaerobic training had no significant effect on endorphin levels, but one session of aerobic exercise significantly increased plasma endorphin levels.

Along with a diminished sensation of pain following a long aerobic exercise session, some people experience a feeling of euphoria and reduced anxiety. This pleasant emotional experience is due to the combined actions of endorphins and the production and release of endogenous marijuana-like compounds; one of the most studied of these marijuana-like molecules is called *anandamide*. Anandamide is manufactured by the body in order to reduce the pain in the muscles and joints that was induced by the extreme exercising. Together, these morphine- and marijuana-like chemicals produced by the body are very effective and also quite addicting. Yes, you can become addicted to molecules produced by your own body. The actions of endorphins and anandamide, as well as a few other neurotransmitters, are capable of inducing you to become addicted to virtually anything, even the repeated painful experiences associated with long-distance running.

The emotional component of this experience, the so-called runner's high, is mostly due to the release of anandamide, the endogenous marijuana-like molecule. The reason is that anandamide can cross the blood–brain barrier, while endorphins produced in the periphery are not able to cross the blood–brain barrier. Fortunately, your brain is able to produce its own endorphins in response to your extreme exercising. The pain reduction depends on the stimulation of cannabinoid (relating to where marijuana acts) and endorphin receptors scattered throughout the body and spinal cord. Inside the brain, after an extreme level of exercising, the decreased pain perception, reduced anxiety, and profound euphoria are due to the actions of endorphin and anandamide produced by the brain as well as two other marijuana-like molecules, 2-arachidonoyl glycerol and N-palmitoylethanolamine. As you can tell, the brain evolved a cornucopia of psychoactive drugs to self-administer in response to the repeated injury that was associated with the risks and tasks of daily living. Long ago, it was quite dangerous to go out and forage for food and mates.

Endurance athletes become addicted to their own endogenous morphine- and marijuana-like chemicals. If you are an endurance athlete, the addictive proclivities of these chemicals, and their ability to produce euphoria and mask your pain, make it highly likely that your brain will insist that you run again as soon as possible. That craving you are feeling for the experience associated with another long-distance run fits exactly into the textbook definition of an addiction. Heather D'Angelo, a neuroscience colleague, vividly described her personal running experience as follows:

> When I run, I am sometimes able to reach this moment of overwhelming freedom and clarity. My body buzzes with excitement and my mind goes numb. It's as though I'm finally freed from the prison of my own mind and all I care about is moving one foot in front of the other. All of my problems feel silly and insignificant. There's no experience quite like bringing myself to the absolute limit—right up to the threshold of what I can endure—and then doing more. The act of redefining my personal limits is thrilling and empowering beyond words. Every time I run, I'm always chasing that moment of absolute freedom from gravity—I'm flying.

If this is an addiction, then it is one that she clearly enjoys. What's going on inside her brain when these feelings occur?

In order to answer this question, a recent study used MRI on humans to monitor changes in activity within the brain following a two-hour run. Surprisingly, this study reported that the endorphin-induced euphoria actually developed prior to running! The authors concluded that prior to, during, and after endurance exercising, the brain has learned to anticipate the likelihood of injury and decides to self-administer a dose of painkillers and euphoria-inducing drugs. This action by the brain is typical of many other well-studied phenomena; expectation of an upcoming event can activate several brain and body systems prior to an event. This response is also part of what is called the *flight-or-fight response* if a dangerous situation is anticipated.

The release of endorphins and anandamide is also a vital component of what is called the *placebo effect*. Humans are extremely adept at self-inducing both euphoria and analgesia, the core consequences of the runner's high, via the placebo effect. Experienced drug users demonstrate a very similar phenomenon; they become euphoric when they first see the needle or the pill, even before they self-administer their drug of choice.

Another great example of a pain-mitigating ability of the placebo effect is the consequence of listening to music while exercising. Listening to enjoyable music while exercising reduces how you perceive your level of exertion, increases your enjoyment of exercising, and enhances your performance; these benefits are primarily experienced only when performing low- to moderate-intensity exercises. Unfortunately, the placebo effect can make the experience of pain more or less agonizing depending on how you feel—for example, if you are fatigued, anxious, fearful, or bored, then the pain associated with exercising becomes unpleasant and intolerable.

When it comes to alternative medicines and therapies that claim to enhance your brain function, never underestimate the power of your own expectations and the placebo effect. Therefore, the best approach, and the cheapest one by far, is to expect great things of your brain and generate your own placebo effect. Much has been written about the positive value of the placebo effect in the practice of medicine, but how this effect emerges and whether it can be controlled are issues that are not yet fully understood. Essentially, scientists have analyzed the effect based on results of placebo-controlled studies of actual drugs on the brain or have compared only the effects of a placebo against the consequences of no treatment at all. Their findings have been intriguing, if still largely inconclusive.

However, in one area of study that is not directly related to an actual treatment, the findings are more definitive. Numerous meta-analyses (which are later analyses of other researchers' data) have shown that the perception of pain can be significantly influenced by your thoughts. In a study published in late 2008, scientists directly investigated the placebo effect on pain perception using a novel test. They measured pain perception in two groups of people—devout practicing Catholics and professed atheists and agnostics—while they viewed an image of the Virgin Mary or the painting *Lady with an Ermine*, by Leonardo da Vinci. The devout Catholics perceived electrical pulses to their hands as being less painful when they looked at the Virgin Mary than when they looked at the da Vinci work. In contrast, the atheists and agnostics derived no pain relief while viewing either picture. MRI scans demonstrated that the Catholics' pain relief was associated with greatly increased brain activity in their right ventrolateral prefrontal cortex. This brain region is believed to be involved in controlling our emotional response to sensory stimuli, such as pain. Perhaps this study has, in fact, shown us the location of the placebo effect.

Other studies using brain-imaging techniques to show correlations between brain activity and the extent of reported placebo effects have demonstrated that some people show greater placebo responses than others, but that everyone appears to be capable of having such a response. There is also increasing proof that the use of placebos might benefit people with Parkinson's disease, depression, and anxiety. In the future, with better testing measures, scientists will likely demonstrate how the placebo effect influences many aspects of our health. In short, the placebo effect is real; we simply do not understand entirely how it works, but the evidence thus far is truly remarkable, particularly with regard to pain. Some people are able to block incoming pain signals or alter how they are perceived. In addition, without a doubt, your mind can make the experience of pain more or less agonizing depending on how you feel—for example, are you fatigued, anxious, fearful, or bored? Do you expect more painful experiences to be coming soon?

There is growing evidence that the individual's genetic makeup, now referred to as the *placebome*, influences clinical outcomes and potentially may allow for identification of patients who might respond to placebo therapies and not require actual medications, with their often-unpleasant side effects for the brain and body. The placebome may explain why some endurance runners get the high faster than do other runners. The genes for four different neurotransmitters have been implicated in the placebo effect. There are at least 10 genes that influence both dopamine and serotonin function that may underlie the robustly demonstrated placebo effect on mood, particularly the symptoms of depression. In addition, there are specific genes that influence the endogenous opiate and cannabinoid neurotransmitter systems that likely underlie the ability of placebos to produce analgesia, as described earlier.

Overall, the pain-reducing action of a placebo seems to be responsible for about half of the response to pain-reducing medications. This is true regardless of whether the active medication is aspirin or morphine, meaning that placebo morphine is significantly more powerful than placebo aspirin. Why? Because people expect that morphine is more powerful than aspirin. There is a downside to this effect: If you do not believe that your medication is going to help you feel less pain, then it probably won't. Negative thoughts and expectations can make pain and depression feel much worse: This is called the *nocebo* effect.

With regard to depression, the contribution of the placebo effect ranges from about 30% to 40% of the response to any standard antidepressant medication. There is one key aspect of the placebo effect that always unmasks its presence in any treatment: Placebo effects always occur faster than "real" drug effects. If you ever read about or hear a claim that the benefits of any treatment are "immediate," you can be assured that it's all due to the placebo effect; so save your money. Reiki therapy is a perfect example of the placebo effect; Reiki "masters" always claim that the benefits are almost immediate. Furthermore, placebo effects do not demonstrate normal extinction. This means that people will continue to claim benefit from a placebo pill even when the pill no longer produces its original benefits. True believers refuse to give up false hope; they just change the nature of their expectation of their placebo.

The placebo effect in humans is fascinating to study. It is influenced by so many different factors. For example, the color of the pill you take influences your expectation of what it will do to you. Obviously, pills can be made any color, yet most people like their antianxiety pills to be blue or pink or some other soft, warm color; they prefer their powerful anticancer pills to be red or brightly colored. Americans do not like black or brown pills, in contrast to the preference of people in the United Kingdom or Europe. Look around the next time you are visiting your local pharmacy. Ask the pharmacy staff if they sell any black pills. Almost all the pills that Americans buy over the counter (i.e., nonprescription drugs) are small, white, and round. Big pills, or pills with odd shapes, are also assumed to be more powerful, or more effective, than tiny round pills. When an experimental drug becomes less effective, a simple change in color or shape restores a drug's ability to produce a placebo effect. In addition, sometimes the benefits originate from the pill-taking regimen. For example, you expect that when you are instructed to take a medication only during a full moon, or only every other Thursday, it must be extremely, almost mystically, effective. Herbalists, chiropractors, and other poorly trained practitioners often take advantage of this concept by recommending odd or excessive dosages of peculiar-looking pills or foul-smelling potions. Some chiropractors like to dazzle their patients with electronic equipment that can diagnose illnesses that physicians and physicists are at a loss to explain.

We all want to believe that the pills we consume, or the therapies that we experience, will help us feel and function better. Fortunately, thanks to the phenomenon of the placebo effect, we do sometimes, but only for a while, benefit from even the most bogus of potions and pills. As Tinker Bell said, "You just have to believe!" If all you're getting is a sugar pill, then does it really matter whether you're fooled into believing the lie? Possibly; it depends on the cost of the sugar pills and the risk one assumes by not taking a medicine of proven effectiveness in a timely fashion for a medical condition. The risk of taking substances that merely promise the elusive Holy Grail of long life and good health may be no less dire, depending on the true nature of the "sugar" that's in them.

Never underestimate the power of the placebo effect to influence these two brain functions: mood and pain relief. If someone ever promises you *fast* relief from pain or sadness, they are selling you a placebo. In the future, with better testing measures, scientists will likely demonstrate how the placebo effect—in reality, your own brain activating a selection of its own neural circuits—influences depression and pain relief so effectively. Depression and chronic pain are the two most common neurological problems, and most people experience them at some point during their lives. Although the placebo effect is impressive, and certainly contributes to the reduction in symptoms of chronic pain and depression, it is not sufficiently powerful that it can completely remove the sensation of severe pain or depression. For these, you may require more powerful pharmaceuticals to achieve relief. In the future, knowledge obtained from studying the runner's high may lead to better treatment approaches for chronic pain following brain trauma.

# 16

# EXERCISE FOR THE INJURED BRAIN

Most people who choose to participate in most sports never come close to experiencing the true euphoria of the runner's high; usually, their only goal is merely to improve brain and body health with the hope of enhancing their longevity. There are at least two ironies to this goal. First, as I have already discussed many times, exercise comes with a biological cost: a reduction in longevity due to the subsequent increase in caloric intake and the additional oxygen that is required to access those calories. Second, participation in sports is now seen as one of the major causes of brain injury, particularly among young people.

A recent study by the Mayo Clinic found that one-third of patients' brains showing pathology and evidence of chronic degenerative diseases had participated in contact sports. The popular press has carried numerous stories about retired players of the National Football League who have a threefold increase in their risk of developing depression as well as a variety of worsening cognitive impairments. Indeed, all athletes, especially young adults, exposed to repetitive concussions are at increased risk of developing cognitive deficits.

I have witnessed first-hand the consequences of high school sports-related injuries to the young brains of my college students, particularly the female students, who tend to play sports that do not require wearing protective gear or clothing. Many of them have told me about the traumatic consequences of participating in volleyball, softball, gymnastics, cheerleading, and soccer in high school. Their most common symptoms are chronic pain, usually due to injuries to their lower spine; emotional instability; and impaired academic performance due to repeated head trauma. They complain to me about sleeping problems, depression, and inability to concentrate in class. The complaints of my students are so prevalent that each year I encourage them to volunteer as control subjects in clinical studies being conducted on campus that will allow them to obtain a free MRI scan of their brains. Sadly, every year, their brain scans discover that the consequences of high school sporting accidents are still lurking inside their heads. One of my students was a boxer in high school; his brain was littered with the scars of old injuries that had only partially healed. Another student had spent so many hours doing headstands or downward-facing dog in her yoga class that the blood vessels on the top of her brain had developed serious malformations that placed her at risk of intracerebral bleeding. Please note: The arteries and veins inside your skull did not evolve to tolerate having the entire vasculature of your body hovering above it for prolonged periods of time. Please, do not even think about doing headstands or handstands for prolonged periods of time.

One potential mechanism that benefits the recovery process following brain trauma due to certain sports is called *angiogenesis*, which is the growth of new blood vessels. Exercising on a treadmill can increase the density of blood vessels by inducing the growth of new vessels in the brain. Daily walking on a treadmill was shown to offer additional benefits, such as improved dendritic plasticity, leading to better brain function, in the brain region surrounding an infarct (injury due to loss of blood flow). The angiogenesis induced by exercise is due to the release of the protein VEGF, as well as other proteins. The growth of new blood vessels promotes cerebral blood flow and supplements the delivery of nutrients to exercise-stimulated brain regions, thus promoting further angiogenesis and reducing the extent of the brain damage following a sports injury.

Through these mechanisms exercise improves neuronal survival, regrowth, and the repair of any neurons that become injured during the trauma. Exercise may also reduce one of the most painful and lingering symptoms following injury to the nervous system, neuropathic pain. This condition is typically associated with a stabbing pain, often in the middle of the night, as well as a chronic prickling, tingling, or burning feeling all day. Uncontrolled, neuropathic pain can have negative consequences on brain health, as well as significantly impairing activities of normal daily living.

## Timing matters

The timing and nature of the exercise intervention after brain trauma are important considerations. Chronic versus acute physical activity can have opposite effects on recovery. High-intensity exercise, such as running, may have a negative effect on recovery from neural injury. Moderate levels of exercising are probably most effective when the nervous system is most vulnerable to recovery-inducing interventions. The most vulnerable period for inducing recovery is the time immediately after the initial insult to the brain or nervous system. The type and duration of the exercise that is ideal to support recovery are still unsettled in the available literature.

## Rest versus exercise

For some physical therapists, the most common conservative prescription following mild brain injury is inactivity until the patient is entirely symptom free. This advice is also controversial due to valid concerns that prolonged rest might be counterproductive to a successful or complete recovery. In addition, such advice is nearly impossible for the patient or the caregiver to enforce. Also, prolonged rest often leads to back pain, a loss of physical conditioning, and, most disagreeable for the brain, social isolation.

In order to address this controversy, a recent meta-analysis of 14 different studies conducted during the past 15 years compared the benefits of rest or exercise during a 21-day period following traumatic brain injury. At this juncture in the book, you are probably not going to be surprised by what they discovered in this large analysis of the current literature. The meta-analysis concluded that exercise was modestly, but not statistically significantly, beneficial to the injured brain as compared to inactivity and rest. However, once again, the devil is concealed in the details. For example, the problem in drawing a positive, much less significant, conclusion was partly due to poor compliance by many of the subjects. Some subjects did not exercise according to the prescribed routines. The difficulty in convincing patients with brain trauma to follow instructions rigorously may underlie why most recent and well-controlled large studies have not been able to make strong statements about the value of exercise. Typically, the authors offered generally weak conclusions that exercise can be "helpful" or that exercise is "believed to have beneficial effects" on brain physiology. These are not robust endorsements. Exercise certainly gives the patient the perception that progress in rehabilitation is being made; fortunately, if only due to the influence of anticipation and the placebo effect, this perception is often associated with a real improvement in psychological, if not physical, functioning. Thus, if injured patients are able to perform moderate levels of physical activity, then they should be encouraged to do so. Even if their brain is not helped by the activity, at least they will gain a sense of control over their current condition.

Proof that exercise is beneficial following severe traumatic brain injury is often even more difficult to obtain than it is for mild brain trauma. For example, a study involving eight weeks of aerobic physical exercise in the early phases of rehabilitation of patients with moderate to severe brain injury showed that although the patients were able to participate in the study and tolerate the exercising routine, they showed no significant improvement. There are at least three reasons for this result: The brain injury was too severe; the intervention was not timed to optimize the brain's responsiveness to the exercising; or the intervention did not last long enough to have a significant impact on the patient's outcome. In addition, most patients undergoing rehabilitation for severe brain trauma remain largely inactive or sedentary when not actually undergoing therapy. It is still controversial whether or not it is wise to let the brain rest before challenging it with exercise. A better understanding of the biology of brain injury will lead to the design of more effective therapeutic approaches.

## Mitochondria—again

In the hours, days, and weeks following the initial accident, a series of secondary biochemical changes develop that lead to a progressive degeneration within vulnerable brain regions. Many of these changes are also commonly associated with advanced normal aging and are thus rather well studied. One of the initial changes involves a dysfunction of the mitochondria inside of the neurons of the brain. The injury to the mitochondria leads to oxidative stress and chronic brain inflammation, which leads to an assortment of degenerative diseases, particularly during the years following the sporting event. These three critical events following the traumatic brain injury—that is, loss of normal energy production, oxidative stress, and long-term brain inflammation—underlie the development of seizures, sleep disruption, fatigue, depression, impulsivity, irritability, and cognitive decline. Although no effective drug treatments are currently available to alleviate all of these detrimental biochemical events in the brain, research has advanced sufficiently for us to understand how specific chemicals in the diet can target the negative effects of oxidative stress and inflammation.

# 17

# DIET AND EXERCISE FOLLOWING BRAIN INJURY

The modest benefits of exercise for patients recovering from traumatic brain injury might be augmented by the addition of certain nutrients to the diet. The choice of nutrients has been guided by the improved understanding of the series of changes following injury that occur in the brain that have short- and long-term negative consequences.

A series of recent studies, conducted primarily using animal models, have discovered that adding certain vitamins and minerals to the diet might alleviate some of the long-term consequences of brain injury—just be careful not to take this recommendation too far. Overall, it is never wise to take megadoses of any supplement since many of them can have toxic effects on your liver and kidneys. It can also be a waste of money since most of the contents of these over-the-counter supplements are never absorbed by the body or are quickly destroyed by the liver and then excreted. People who take riboflavin (vitamin B2) supplements usually witness this effect a few hours later when their urine turns bright yellow—that's all of their supplement going down the drain.

The overwhelming evidence currently available today indicates that it is always most effective to obtain nutrients via their natural sources. Unfortunately, almost all of the evidence that supports the claims in the following paragraphs is entirely correlational. Actual interventional studies are rare because they are expensive to perform on humans. Thus, please consider the following nutritional suggestions with a fair dose of skepticism.

Supplementation with vitamins B3 (found in white meat from turkey, chicken, and tuna), D (most dairy products, fatty fish such as salmon, tuna, and mackerel), and E (nuts and seeds, spinach, sweet potatoes) improved cognitive function following repetitive concussive brain injury. Magnesium and zinc are both depleted following traumatic brain injury. Zinc supplementation for four weeks reduced inflammation and neuronal cell death and decreased the symptoms of depression and anxiety in rats following traumatic brain injury. Both zinc and magnesium can be obtained by eating nuts, seeds, tofu, wheat germ, and chocolate.

The omega-3 fatty acids and alpha-linolenic acid were also shown to be neuroprotective in animal studies whether taken prior to or after the injury. Thus, people who participate in contact sports might want to add these fats to their regular diet. However, don't waste your money on alpha-linolenic acid or similar supplements; adequate amounts are easily obtained via a diet containing fatty fish, flaxseeds, canola oil, soybeans, pumpkin seeds, tofu, and walnuts. Sulforaphane was shown to improve blood–brain barrier integrity, reduce cerebral edema, and improve cognition in a rodent model of traumatic brain injury. Sulforaphane can be obtained with a diet containing brussels sprouts, broccoli, cabbage, cauliflower, kale, broccoli sprouts, turnips, and radish. Finally, enzogenol improved cognition when administered to brain-injured patients in a randomized, controlled study. Enzogenol, a water extract of the bark from the Monterey pine tree (*Pinus radiate*), contains high levels of proanthocyanidins. Once again, do not waste your money on bark extracts; proanthocyanidins are easily obtained by consuming grapes (seeds and skins), apples, unsweetened baking chocolate, blueberries, cranberries, bilberries, black currants, hazelnuts, pecans, and pistachios.

Natural antioxidants found in fruits and vegetables, like polyphenols, provide protective effects for the brain through a variety of biological actions. Polyphenols are plant metabolites; the most common dietary polyphenols, and probably most thoroughly investigated, are the flavanols quercetin and catechin. Green tea polyphenols are believed to be strong antioxidants. Tea contains a number of bioactive chemicals and is particularly rich in flavonoids. Epidemiological data indicate that tea drinkers tend to be much skinnier than coffee drinkers; no one has yet determined why. Curcumin, derived from the spice turmeric, the powdered rhizome of the medicinal plant *Curcuma longa*, has been used for many centuries throughout Asia and India as a food additive and a traditional herbal remedy. Curcumin has potent antioxidative and anti-inflammatory proclivities. The best evidence for the beneficial effects of curcumin comes from India, where this spice is consumed in rather high doses virtually every day. What is true for exercise is also true for nutrients: You need to get the right amounts as often as possible.

The non-flavonoid resveratrol is present in red wine and the skin of grapes. These grapes use resveratrol to defend themselves from fungus. Resveratrol is the focus of many current clinical trials and is believed to be partially responsible for what has been termed the *French paradox*. This paradox refers to the claim, now mostly discredited, that France enjoys a relatively low incidence of coronary heart disease despite a diet high in saturated fats. The one critical drawback of resveratrol is that its level in red wine is so low that you would need to consume two cases of wine every day to achieve the benefits seen in animal studies.

The French may also benefit from their classic Mediterranean-style diet. They also consume olive oil instead of unhealthy vegetable or coconut oils. Olive oil, particularly the extra-virgin preparation, contains lots of antioxidants (just look at that rich color; it screams antioxidants) and also has significant anti-inflammatory properties. Everyone knows that these are the two most important properties of healthy foods.

Your body clearly benefits from eating olive oil. Does your brain care? Answering this question is not easy. First of all, olive oil is a complex blend of various chemicals. It would be nice to know which of these is most advantageous and whether they influence brain health directly or indirectly. A recent study determined that olive oil modifies how your gut microbiome communicates with your brain. Olive oil's beneficial effects on the human brain and body are likely related to the presence of the polyphenols hydroxytyrosol (HT) and oleic acid (OA). HT protects cells that are under oxidative stress. OA is a monounsaturated omega-9 fatty acid whose levels are generally higher in olive oil than vegetable fats. It has beneficial effects on blood cholesterol levels. The study discovered that extra-virgin olive oil, OA, and HT reduced the level of various pro-inflammatory proteins in the brain. So, yes, your brain benefits a lot from adding olive oil to your diet, and these benefits originate in response to the changes that olive oil makes to your gut microbiome. Olive oil should become a big part of everyone's diet.

Research on naturally occurring "drugs" has discovered a plethora of beneficial compounds. The polyphenols mentioned earlier are everywhere in nature! More than 50 different plant species and over 8,000 such compounds have been identified in plant extracts. Obviously, investigating the multiple health benefits of these natural chemicals poses an enormous challenge. One of the more promising plant species that might offer numerous health benefits is marijuana. It would be hard to ignore the current excitement regarding the potential benefits of using marijuana. Unfortunately, due to its legal status, carefully controlled interventional studies on the benefits of marijuana have not yet been conducted. A few very small human case studies and a considerable number of animal studies published by my laboratory suggest that marijuana, or some component of the plant, has significant neuroprotective effects. In a three-year retrospective study of patients who had sustained a traumatic brain injury, decreased mortality was reported in individuals who admitted to using marijuana. There is an obvious sampling bias embedded in this study; only those who were alive when the study was conducted could answer the question of whether or not they had used marijuana.

The National Football League and its Players Association recently agreed to consider the use of marijuana as a pain management tool for its players. This decision is long overdue and, in my opinion, the focus on pain management is far too narrow given what is now known about the actions of the various components of the plant. Neuroscientists have collected evidence that the chemicals in marijuana can prevent, or significantly reduce, the short- and long-term degenerative changes that occur following repeated head trauma. These negative changes in brain chemistry and physiology, such as a long-term increase in the level of inflammatory proteins and ROS, underlie the fact that American football players have a threefold increased risk of developing depression, cognitive impairments, and dementia. Some players also report insomnia and nightmares as a consequence of the head trauma. Marijuana has been shown to reliably reduce insomnia and nightmares. However, the reason that the nightmares are reduced is related to the distorted sleep patterns that are induced by marijuana. Unfortunately, marijuana, similar to all of the known sleep aids, reduces the amount of REM sleep you experience each night.

Overall, current evidence supports the safety and efficacy of daily low-dose cannabis vaporization and oral mucosal delivery for the treatment of pain and the reduction of symptoms that develop as a consequence of repeated brain trauma. Due to a lack of adequate clinical evidence from human trials, no one can yet predict the best dosage regimens. However, serious adverse events at these doses are rare and cannabis products are generally well tolerated. My research on animal models suggests that daily low doses of marijuana—one puff per day—are effective in reducing the negative impact of brain inflammation in older brains.

# 18

# EXERCISING THE OLDER BRAIN

Fortunately, we are all getting older. Unfortunately, as we get older our brains fail to function as quickly or efficiently as they did when we were much younger. I have spent my entire research career studying the effects of drugs and nutrients on the aging brain, thus this topic was one I enjoyed investigating. It is also a topic that is a popular target of charlatans and frauds trying to make a quick profit on people who are desperate to reverse the effects of aging on their brain. Therefore, I am going to begin by discussing interventions that do *not* help the aging brain and then later investigate whether exercise falls into this same category.

The decline in cognitive function that occurs with normal aging is so bothersome to most of us that we frequently turn to unproven claims by charlatans who hope to make few dollars on our desperation. Here's how it works: Unethical purveyors of these products find a condition associated with normal aging for which the medical establishment offers no therapies (which is, let's face it, almost all of them!). They propose a product that treats this condition after searching the "scientific" literature for some natural substance that provides some smidgen of possible help, based on unreliable and often biased evidence. If they can link it to use by the ancient Chinese or Native Americans, so much the better. They use shrewd language to make a mountain out of a molehill, and blow some scientific-sounding mumbo-jumbo around and feature the words "natural" or "neuro" prominently in their advertising. They deliberately choose a condition that makes it difficult to determine whether their product is effective, such as during the early stages of dementia. They all offer a money-back guarantee, but when their customer wants to exercise this right, they will ask for evidence that the product did not work. Since it is logically impossible to prove a negative effect, the customer cannot provide such evidence, and the charlatan keeps the money.

The sad truth is that no single molecule or formula can correct the damage produced by the complex and numerous biological processes that underlie normal aging. It is not that simple.

## Snake oil science

The best advice today for your aging brain is this: Avoid the snake oil. The internet is full of misleading information about so many things related to getting old. It is very easy to purchase lots of different "cures" for mental and physical decline. I truly wish that a cure did exist; I'd be first in line to get it. We would all prefer to defy the aging process by simply taking a pill. Sadly, no such cure exists. The fact that science has not yet invented a true brain enhancer has not stopped people from selling drugs, ancient elixirs, unusual therapies with mystical names, and hundreds of books that all boast of the properties of this or that miracle, age-defying brain booster. If someone might gain financially from your gullibility, then what he or she is selling is probably useless; furthermore, there is no guarantee that it is safe.

Nothing has yet been discovered that can significantly enhance cognition or prevent brain aging. In spite of this, someone will always be willing to sell you a pill containing something exotic and completely useless, such as Prevagen. Apparently, we are still as vulnerable to the same false claims that sold similar magical elixirs to our grandparents and their grandparents. The Federal Trade Commission and others have argued that Prevagen does not show any significant improvement in the treatment group over the placebo group. Indeed, Prevagen actually performs significantly *worse* than a placebo if you take it for more than two months.

Why do so many people fall under the spell of charlatans? How can so many people feel so strongly that these drugs work on them? The answer is quite easy to summarize in three little words—the placebo effect. Essentially, we want these drugs to do something, anything, so we fool ourselves into thinking that they do. After all, you've just spent a lot of money on this pill!

### Worthless multivitamins

Americans spend billions of dollars on multivitamin–mineral supplements, but are they getting a return on their investment? Not always. Numerous studies involving hundreds of thousands of men and women of all ages and genetic backgrounds have found little or no long-lasting benefits from taking a daily multivitamin–mineral supplement. One large study followed the lives of 182,000 men and women; those who took daily multivitamins did not live longer or have less heart disease or cancer, the two primary killers of Americans. In a study following 161,000 postmenopausal women for the effects of multivitamin use, those who took daily multivitamins were not less likely to develop breast, ovarian, or any other cancer than women who did not take multivitamins. Among 83,000 middle-aged to elderly men who were followed, those who took daily multivitamins were no less likely to die from coronary heart disease or stroke than were men who did not take multivitamins.

So maybe you are still hoping that taking multivitamins, at the very least, might offer some modest benefits for your substantial investment of time and money. Sadly, after many decades of research, the evidence for even modest health benefits is still very iffy. Even when researchers examined the benefits of multivitamins for symptoms associated with the common cold, they could find no significant health benefits. What about the brain? Surely multivitamins help the brain; just look at the hundreds of claims on the internet! Nope.

When healthy older men and women, aged 60 to 91 years, were given daily multivitamin supplements for six months, they demonstrated no significant improvements on memory or other cognitive function tests as compared to being given a sugar pill. Vitamin E supplements are no longer recommended for brain health; indeed, the high doses originally thought to help slow the onset of dementia with aging are now recognized to increase the risk of cerebral hemorrhage. In spite of a total lack of evidence that multivitamins offer any real long-term health benefits, we are all addicted to them. This is why Americans produce the most expensive urine in the world; we merely excrete whatever our bodies do not need immediately. Our bodies do not store most vitamins for later use.

Concerns about vitamins and the balance of their risks versus their benefits were expressed during the early 1950s when parents learned that powerful chemicals—vitamins and minerals—were being added to their child's favorite breakfast cereals. The solution for some manufacturers was to offer the pills as effigies of popular cartoon characters. As recently as 2004, Denmark outlawed some vitamin-fortified cereals because of concerns that extremely high levels of vitamin B6, calcium, folic acid, and iron might achieve toxic levels if eaten daily; the risk is particularly high for young children—the principal consumers of many enriched cereals. The Danes might be overreacting; however, it's probably not a good idea to take a daily multivitamin if you're eating a cereal containing 100% of the daily recommended levels.

Overall, most of us are wasting our money because we have been completely sold on the belief that we need these chemicals to be healthy. The epidemiological evidence does not support this belief. Indeed, some things that were once thought to be critical for our good health, such as selenium or vitamin A, are simply not as big of a concern; indeed, recent recommendations are to avoid high doses of these two supplements.

However, and this cannot be overstated, some people *do* need supplements of certain vitamins and minerals, either because of poor diets, disease states, advanced age, gender, or the lack of sunshine, to name just a few of the major reasons. We are all aware of the scientifically supported recommendations that some of us should consume more vitamin D (particularly if you are taking statins to reduce blood lipids), take folic acid during pregnancy, take certain B vitamins to maintain mental and physical health, and consume additional iron, particularly if you're a woman over 50 years of age.

There are almost as many nondrug interventions for your brain, and they also lack any shred of scientific proof. These interventions usually invoke the actions of some mystical force that physicists have repeatedly failed to discover. The fact that these interventions lack any scientific support does not deter desperate people, usually encouraged by the willfully ignorant who seek them out. Craniosacral therapy, ear candling, magnet therapy, crystal healing, cupping, Rolfing, neurolinguistic programming, psychokinesis, and primal therapy are just a few of the frequently mentioned examples of completely ineffective interventions that are wholly lacking in any scientific support. In addition, "energy medicines," such as Reiki, which involve "laying on of the hands," or various types of hand waving above the body, or any of the numerous naturopathic practices, have never been proven to provide any medical relief beyond that produced by the placebo effect. Always remember, placebos are very effective for pain relief and mood enhancement, the two primary reasons that people turn to these irrational interventions. In addition, the practitioners of these absurd healing methods always claim that the benefits appear quickly; this is yet another key feature of the placebo effect. Placebos act quickly; medically effective therapies never do. Because real medications that produce real benefits take so long to work, people often give up too soon and turn to alternative therapies.

Interventional studies using natural antioxidants and anti-inflammatories in the diet have become attractive options for patients with mild sports-related injuries or brain injury. Indeed, lifelong modifications of the diet may be essential to maintaining whatever level of recovery you achieved after the injury. With normal aging, your brain's ability to compensate for the injuries sustained during youth and adulthood begins to weaken and it becomes harder to recover lost abilities. Young brains can compensate and mask underlying injuries following a sporting accident. Old brains do not compensate as well and may fail to mask the cognitive consequences of injury, particularly when fatigued or under the influence of brain depressants such as alcohol. So, can exercise help the older brain feel younger?

### Exercising the older brain

In the same fashion that the internet is full of claims about the benefits of nonsense therapies for the aging brain, the internet is also full of hype and hyperbole about the benefits of exercise for older adults. Thus, it is difficult to discern whether any relationship exists between exercising and improved cognitive performance. If you have any hope of motivating your elderly loved ones to exercise, it would be useful to have positive evidence from reliable scientific studies to support your pleas. Without doubt, you can confidently tell your elderly relative or friend that regular exercise has been found to help prevent heart disease, diabetes, and certain forms of cancer; for arthritis patients, it can also reduce pain and might allow them to take less medication each day. Regular exercise, even a 30-minute walk every day, can improve balance, flexibility, endurance, and lower body strength. Taken together, regular exercise will likely help your elderly loved ones to live a longer and healthier life.

Given all of this positive evidence, you might predict that the benefits of regular exercise should translate into a healthier brain as well. Unfortunately, in spite of the fact that the current literature on the topic contains more than 1,000 clinical trial reports, almost 200 reviews, and 50 meta-analyses, there is still no practical or detailed guidance on how to utilize exercise to improve brain health in older adults. No one has yet defined the exercise dose that is either sufficient or optimal to induce cognitive benefits in the elderly. Overall, the effectiveness of exercise in older adults has been rated by the authors of these numerous large and small studies as ranging from only slight to moderately beneficial, respectively. Typically, small studies tend to report greater positive effects than are usually reported by larger studies because larger studies take into account more of the natural variance found in larger populations. Smaller studies can provide useful insights but have limited construct validity.

Given the very limited effectiveness of the available treatments for cognitive decline or dementia, it would be beneficial to discover an effective method to promote healthy brain aging. Overall, the current evidence does not support the claim that exercise can counteract the diverse and complex effects of aging on the brain. There are probably numerous reasons that older adults do not benefit significantly from exercise as compared to young people. For example, aged athletes usually exercise at a lower absolute intensity, making it difficult to detect a significant benefit using current methods of investigation. The benefits of exercise in older adults may be real but often too subtle to notice. In addition, older adults display lower rates of carbohydrate oxidation, a chemical process that is critical to maintain an active brain, when exercising at the same relative intensity as younger adults.

A few years ago, a small study investigated whether long-term dance experience improved cognitive performance and gray matter brain volume in women aged 65 to 82 years. The control group were non-sedentary people without any dancing experience, who were similar in age, education, IQ score, lifestyle and general health factors, as well as fitness level. (Sometimes, finding the ideal control group can be quite difficult.) The authors reported no benefits in any of the cognitive domains tested, including executive control, perceptual speed, short-term memory, and long-term memory. No dancing-induced changes in volume were observed in the hippocampus. Dancing can be fun as well as emotionally rewarding for older adults, but they should not expect dramatic benefits for their brain.

A recent massive study published in a prestigious clinical neurology journal collected data from over 11,000 males and females and grouped them according to three different cognitive levels: older healthy adults, individuals with mild cognitive impairment, and individuals with dementia. The average age was 73 years and most of the participants reported that they led a sedentary lifestyle prior to enrolling in the study. The authors concluded that exercising was associated with improved cognitive performance in older adults and that this benefit was observed independent of whether the elderly person demonstrated cognitive impairment at the beginning of the study. Sounds good thus far; it's consistent with what everyone expects to hear. Once again, though, the devil is hiding in the details and the actual data are not as convincing as you might wish. For example, 53 of the studies that they analyzed had a known high-risk bias because neither the participants nor the testers were blinded to the study design. The people who were responsible for reporting the outcomes of the exercise by the older adults were also responsible for directing the exercise routines. In fact, over half of all the 11,000 participants were aware of what was going on in the study and the results that were expected from them. This situation allows for a considerable amount of bias that completely undermined the value of the studies.

The correct interpretation of these studies was made more difficult because the duration of the exercising varied between

10 and 100 hours, and ranged from four to 44 weeks, for about 90% of the participants. Some of the studies utilized high-intensity exercising, while others utilized medium- or light-intensity exercising. Unfortunately, 20% of the studies did not even bother to report the exercise intensity. Thus, the inconsistency of the conditions of the various studies contributed to a significant amount of variation in the outcome.

Overall, how long each older person exercised, how often they exercised per week, or the duration of the exercising did not correlate significantly with improved memory or any other cognitive function. About half of the cognitive measures were improved with exercising, while the other measures were not. Too often, at this point in the analysis, many authors would cherry pick the positive outcomes and publish those. However, even the modest positive outcomes require numerous caveats. The authors grouped the varied psychological tests that were described in the many different publications they reviewed into multiple categories: executive functioning (the ability to plan, organize, and complete tasks), thinking speed, the ability to pay attention, working (short-term) memory, and the ability to judge distance and navigate around in space. The problem is that no one is certain about the validity of these broad cognitive domains for judging the benefits of exercise. Consequently, it is nearly impossible to conclude that exercising was in fact the cause of any cognitive improvements.

Given the difficulties in conducting this type of analysis, it is not surprising that the study reported that exercise did not benefit patients with mild cognitive impairments with regard to almost all of these cognitive measures. Overall, this very large study found no relationship between improved cognition and how long the patients exercised each week. Essentially, for most elderly humans, regular exercising will offer only modest cognitive benefits. Furthermore, the more cognitive decline patients exhibit when they initiate an exercise routine, the less likely it is for exercise to provide a positive benefit to their brain health or cognitive abilities.

One day, due to the collective consequences of being alive for a very long time, and advanced by your daily habit of eating and breathing, you will experience a decline in physical or cognitive health that may require extended bedrest and prolonged physical inactivity. The lack of activity induces negative functional alterations in many different organs, including skeletal muscles. Because the relationship between physical activity and body health has been so well characterized, it is not surprising that involuntary inactivity can have a significant negative impact on the hormonal signaling by muscles that indirectly influences brain function, such as neurogenesis and improved blood flow.

Sometimes the inactivity is not involuntary but rather the nature of the job. For example, consider the challenge of spending long periods of time moving around in zero gravity. Astronauts contract their muscles and manipulate objects to move themselves around in their spaceship; however, these movements require very little muscular effort. Their muscles are not inactive; rather, they contract their muscles against minimal resistance—this is very different from what happens on Earth. Studies performed on astronauts since the 1960s have taught NASA quite a lot about the consequences of inactivity and space travel on both muscle and brain function. These studies have determined that longer stays in space result in prolonged muscle deprivation that may negatively affect neurogenesis and the interaction between neurons and muscle cells. Many of these changes resemble those associated with normal aging. For example, zero gravity has consequences on muscle chemistry that indirectly influence the brain via changes in blood levels of pro-inflammatory proteins. The neuroinflammation impacts brain function and could have significant consequences for cognitive function and general brain aging during long-duration space flights. Recent studies on animals have shown that male brains are more vulnerable to the consequences of space travel than are female brains. Compared to females, males develop more severe levels of brain inflammation that impair their cognitive abilities. These data suggest that female astronauts might be able to tolerate the rigors of long-term space travel far better than male astronauts.

Physical inactivity, obesity, brain injury, and inflammation are all risk factors that determine how well we age cognitively. Exercise may directly oppose the effects of these risk factors. However, the amount of physical activity necessary to provide protection from these risk factors is unknown. The protective impact of exercise on degenerative disorders of the brain associated with normal aging is still under evaluation, since the mechanisms underlying the potential benefits have not been determined yet. This is due to the heterogeneity of the causes and the course of degeneration in pathologies affecting middle-aged and older adults such as Parkinson's disease and Alzheimer's disease, as well as devastating diseases that have an earlier onset such as amyotrophic lateral sclerosis. For Parkinson's disease and Alzheimer's disease, pharmacological therapies are mostly palliative—they treat the symptoms without influencing the underlying mechanisms; thus, they are not curing and often produce only very modest benefits, often with significant and unpleasant consequences, for the patients. There is growing evidence suggesting that regular physical activity, if started a few decades before any symptoms appear, can generally slow down aging and prevent the onset of degenerative diseases. This discovery is consistent with the conclusions drawn from studies of other lifestyle risk factors. The earlier that people change their lifestyle risks, including obesity, poor diet, inactivity, and smoking, the greater the beneficial consequences on brain and body health.

Inactivity, in particular, increases the risk of falling for older adults. As we age, the risk of falls increases. A fall can produce significant health problems for the elderly and accelerate cognitive decline. Falls may be becoming more lethal; more than 25,000 older adults (defined as older than 75 years) died from a fall in 2016. Some forms of exercise might help mitigate the risk of falling. Taking a daily walk is enjoyable and beneficial for the heart and lungs, but one of the best exercises for preventing falls is Tai Chi. Tai Chi offers postural training and improves stability so that you can catch yourself before falling. Tai Chi has been found to improve balance and motor control in patients with Parkinson's disease.

Parkinson's disease is a neurodegenerative disease characterized by resting tremors, rigid muscles, and limited freedom of movement. A meta-analysis of data collected from over 500,000 adults who participated in eight different studies over a period of six to 22 years discovered that men who reported doing regular exercising as young to middle-aged adults were 32% less likely to get Parkinson's disease. That's an impressive effect and appears to contradict a point I raised many times in previous chapters about the link of oxidative stress, exercising, and various neurodegenerative diseases, such as Parkinson's disease. Once again, the devil is in the details. Unfortunately, the effect was observed in men only; regular exercising did not reduce the risk of getting Parkinson's disease for women. These correlational results should raise significant doubts in your mind about what these studies are actually demonstrating. What else were these men doing? Do men who exercise regularly exhibit other behaviors that might prevent Parkinson's disease? Could these other, non-exercising, behaviors have contributed to the apparent protection these men received? For instance, did these men drink a lot of coffee?

Recently, a series of studies demonstrated conclusively that drinking five cups of regular coffee every day for five years reduces the risk of Parkinson's disease by almost 85%! Coffee drinking was almost three times more effective than exercising. Unfortunately, similar to what was observed for exercising, women did not obtain a similar benefit from drinking coffee. One conclusion from these correlational studies would be that women should consider giving up both coffee and exercising; it is not going to help them much as far as brain aging is concerned. Such a conclusion does follow from the available data but seems completely silly to me. Once again, correlational findings should never be trusted until the underlying biological mechanisms are well understood.

Some correlations seem entirely plausible; others do not. For example, the incidence of shark attacks on humans correlates very highly with ice cream sales. Also, during the past 10 years, the increasing incidence of injuries from having a television fall on someone correlated significantly with the enrollment of undergraduate students in American universities. You might be able to find a reasonable explanation for the first example: People eat more ice cream during the summertime while they are enjoying a sunny day at the ocean, where sharks live. I have no idea how to explain the increasing danger of falling televisions to young people eager for an education. In contrast to these questionable correlations, the benefits of coffee are now known to be mechanistically linked to the actions of caffeine within the brain. Caffeine, whether from coffee, tea, or colas, is equally effective. No one knows why caffeine is not as protective in the brains of women, but it may have something to do with the complex actions of ovarian hormones.

Frequently, finding an explanation for the causal relationships discovered in epidemiological studies requires careful studies using valid animal models, if they exist. For example, moderate exercise training exerted neuroprotective effects in a rat model of Parkinson's disease, enhancing neurogenesis associated with increased BDNF and improvement in movement. Although it is not clear how much these animal models relate to Parkinson's disease, the experimental studies suggest that neuroprotection and plasticity are stimulated by exercise in cortical circuits to delay the onset of degeneration. So, while exercise overall is not directly slowing down the aging processes in the brain, exercise may contribute to the protective processes that reduce the impact of brain injury or degeneration. The next section discusses the multiple molecules that respond to exercise and might benefit brain function directly or indirectly.

Aging-induced cognitive dysfunction is associated with a decrease in hippocampal expression of PGC-1alpha, irisin, and BDNF—three important molecules in the body whose levels are increased by exercise in young adults. However, no evidence is available that actually proves that exercise improves cognitive functioning via production of these proteins. Irisin is capable of increasing BDNF levels in the brain; thus, this protein might underlie some of the benefits on cognition. Irisin levels are reduced in the hippocampus of humans with Alzheimer's disease. However, so many different proteins, hormones, and neurotransmitters are affected in the brains of patients with Alzheimer's disease that it is impossible to assign causality to the reduction in any one of them.

BDNF is not the only hormone that is decreased in the hippocampus of elderly, memory-impaired humans. Decreased levels of two other important hormones, VEGF and IGF-1, are also associated with age-related hippocampal dysfunction and memory impairment. VEGF is hypoxia-inducible, which means that it responds whenever oxygen levels in the blood are too low. VEGF promotes the formation and growth of blood vessels for the obvious purpose of bringing additional oxygen-rich blood to specific regions of the brain. Increased blood flow can support additional brain activity and is often associated with improved brain function. VEGF is produced by muscles, is released with exercise, and is able to cross the blood–brain barrier to support the growth of new neurons and new blood vessels to support the activity of newly formed neurons. VEGF exerts a pivotal physiological role in the regulation of BDNF production, a role that may underlie how muscles induce the brain to produce BDNF.

IGF-1 is a hormone similar in molecular structure to insulin. It is released by the liver and is mainly known for its role in energy metabolism and homeostasis. Thus, this hormone has a broad range of effects on brain and body health. IGF-1 can cross the blood–brain barrier; in the brain it can modulate synaptic plasticity and synapse density. Age-related reductions in

IGF-1 have been associated with decreased cerebrovascular density and blood flow. Reductions in the constant supply of oxygen to the brain could underlie aspects of age-related cognitive impairments. Studies in animals suggest that exercise protects the brain from damage through increased uptake of circulating IGF-1 into the brain. In contrast, recent findings regarding the role of IGF-1 in the brain of exercising humans are not as convincing and are often contradictory. Earlier studies seriously exaggerated the role of IGF-1 in exercise-related changes in brain function. A recent review of seven large studies found that exercise induced an increase in blood levels of IGF-1 in three studies, no changes in IGF-1 levels in three studies, and a significant reduction of IGF-1 in another study. In none of these studies did increased blood levels of IGF-1 lead to improved cognition.

Studies on humans and experimental animal models have provided converging evidence that aerobic exercise does not enhance serum levels of BDNF, VEGF, and IGF-1 in healthy elderly subjects. Although a three-month intervention with regular physical activity significantly increased fitness levels in the elderly participants, there was no effect on BDNF, VEGF, or IGF-1 levels in the blood or changes in hippocampal blood flow. The benefit from regular exercise on cognition in elderly humans, if any occurs, might be related to its anti-inflammatory effects rather than changes in these three proteins.

The benefits of exercise appear to be inverted for amyotrophic lateral sclerosis, also known as Lou Gehrig's disease, particularly when physical activity is intense. A recent analysis of data from thousands of professional Italian football (soccer) players established that the rates of morbidity were significantly increased for amyotrophic lateral sclerosis; those at the highest risk were the footballers who played for more than five years. Other studies showed a higher risk of amyotrophic lateral sclerosis among long-distance runners and rugby players. One recent study reported an inverse correlation between body mass index, a measure of how overweight someone is, and the risk of developing amyotrophic lateral sclerosis. This is an astonishing finding because it implies that being physically *unfit* prevents the onset of this devastating neurological disease. Obviously, some truly interesting discoveries remain to be made about the connection between exercising and amyotrophic lateral sclerosis.

### Environmental enrichment is beneficial

If exercise does not significantly benefit the aging brain, what does? Recent studies of humans and animals demonstrate that environmental enrichment can enhance neuroplasticity, increase expression of growth factors, and enhance hippocampal neurogenesis. Environmental enrichment is defined as the addition of social, cognitive, and sensory stimulation. MRI scans in human subjects showed that an enriched environment induced region-specific changes in gray matter volume and partially rescued age-related reductions in cerebral blood flow. These findings demonstrate that sensory enrichment alone, without physical activity, can ameliorate many features typical of the aging brain by inducing many of the same changes in the brain that are produced by regular modest physical exercise in young adults.

Recent studies using animals have confirmed the results seen in human studies. One interesting study, published in late 2019, compared the effects of exercise with environmental enrichment on the cognitive abilities of aged mice. Environmental enrichment reversed the negative effects of aging on anxiety and memory. In contrast, exercise produced no benefits on either of these measures. The best advice one might draw from these studies is that you would be well served to take a long walk in the woods.

One recent study investigated "exergaming," a combination of physical exercise and simultaneous cognitive stimulation, to determine whether both together might be more beneficial than either alone in cognitively impaired older adults.

Individuals with mild dementia were randomized and individually trained three times a week for three months. The results of this interventional study demonstrated that although psychomotor speed improved with exergaming, there was no significant improvement observed for executive functioning or memory abilities. I would not recommend wasting your time and money on exergaming.

### Treatments not worth the time or money

There are plenty of other pseudoscience ideas that people like to invest time, effort, and lots of money into based on dubious claims. Transcranial direct-current stimulation has become an increasingly popular way to manipulate brain activity in order to enhance cognitive and physical performance and endurance regardless of age. The idea is to apply a mild electrical current to the brain; typically, a dose of about 1 to 2 milliamps is applied to the scalp for about 30 minutes each day. You can actually feel the stimulation because it produces a tingling or mild stinging at the site of the electrode. This tingling contributes to the placebo effect, which supports the subject's impression that the stimulation is working. The expectation by the patient is that this electrical stimulation will result in changes in the activity of neurons and an alteration in brain chemistry.

This all sounds quite reasonable; however, the success of the method depends on getting a sufficient amount of electricity across the scalp and skull in order to reach the brain floating below the electrode. Unfortunately, this is not likely to occur. Most of the electrical current that attempts to pass into the brain is actually shunted through the scalp and does not penetrate the skull. A small amount of current does penetrate into the upper layers of cortex, but at this point it is too weak (probably less than 120 microvolts) to trigger the action of more than a few superficial neurons. A systematic examination of the 420 articles published on the topic discovered that the benefits of transcranial direct-current stimulation were only reported when the number of subjects in the study was small; the benefits disappeared when large numbers of subjects participated in the studies. This is typical for most pseudoscience techniques: The benefits vanish in larger studies. Overall, the benefits of transcranial direct-current stimulation on physical performance are small or nonexistent; furthermore, the current literature is most likely biased by low-quality studies and the selective publication of significant results. Best not to waste your money on this pseudoscience gimmick. However, it probably *is* worth spending your money on a new pair of walking shoes.

### Exercising and creativity

> All truly great thoughts are conceived by walking.
> Friedrich Nietzsche

Walking has been reported to enhance many different cognitive abilities, including one of the most elusive—creativity. Writers such as Henry David Thoreau, J. K. Rowling, and Ernest Hemingway have claimed that walking is a reliable cure for writer's block. Is this true? Does exercise, even in this most modest form, directly influence our brains to create new ideas and discover novel insights to our problems? This potentially important question requires an answer, and scientists have made numerous attempts to provide a detailed psychological or neurological mechanism.

First of all, it is very difficult to study the creative process. The first challenge is to define it so that all investigators agree on what is being studied. Creativity is often described as the ability to produce novel and original ideas. Judging what is deemed creative can still lead to contradictory conclusions across studies. In 2014, a group of psychologists conducted studies on young college students to investigate a potential explanation. They discovered that walking significantly enhanced creativity in virtually every person tested. Participants who walked were always far more creative than those who just sat for a similar period of time. Interestingly, it did not seem to matter whether the walkers walked outdoors or walked indoors on a treadmill. I was surprised that the great outdoors did not convey some added creative units; I guess Thoreau could have saved some effort and just paced back and forth in his living room.

These walking studies were carefully designed to eliminate potential alternative explanations. For example, they ruled out the contribution of simple leg movements and the potential stimulation of the environment, such as being outdoors or indoors. Interestingly, walking was not actually required; dancing produced similar results (except in the study of elderly women). Was walking truly effective? One of the drawbacks of this and many earlier studies was what the psychologists instructed their control subjects to do while the test subjects were out walking. The controls sat still and did nothing. Is that really the most appropriate "control" behavior? No, it is not. In order to understand why, let's look at what else stimulates the creative process.

Walking is repetitive and often unconsciously controlled. If you are walking a familiar trail, your mind tends to wander away from thinking about what you are doing and becomes focused on a random string of thoughts. Your brain goes on a mental hiatus and just experiences the rhythmic up-and-down movements of its visual images and regular rhythms from your heart and lungs. Walkers claim that the transition into this mental state is necessary for the creative thoughts to materialize. What is the critical component of walking that induces this transition from random thinking into creative thinking? Some psychologists believe that random thinking, or chaotic thinking, also gives birth to creativity. In order to offer a potential answer, consider what drives the creative process for creative people who do not walk. Mozart wrote that he often created his best music "when I am traveling in a carriage." He kept a notebook with him in his carriage to jot down ideas and melodies. Some successful authors report that driving a car helps with writer's block. According to equine therapists, riding horses also increases creativity.

Apparently, moving through your environment is also beneficial even when you are not the one doing the moving, given that people become creative while riding in a carriage, on a horse, or in a car, or simply rocking in a chair. Is creativity initiated by the rhythmic sensory stimulation of the brain, or is muscle contraction necessary and sufficient? Applying transcranial direct-current stimulation to the brain of reclining inactive subjects in order to increase neuronal excitability in those areas related to creativity is claimed to enhance the creative process. This would be an exciting claim if anyone knew what areas of the brain are responsible for creativity, and if transcranial direct-current stimulation could be proven to be more than just another form of pseudoscience. If these claims and the conclusions of recent studies are true, then exercise, walking in particular, is not required in order to induce creativity. All you may need to do is take a ride in a car or on a horse. Put a hold on buying those new walking shoes for now.

One recent study investigated the effects of exercise on creativity in children. A group of young students either walked on a treadmill, listened to music, or sat quietly. They then completed four creativity assessments: idea fluency, cognitive flexibility, conceptual originality, and the ability to produce elaborate stories quickly. Exercise improved none of these measures of creativity. Although the boys had better physical fitness, they were not more creative than the girls. The confusion in the literature probably has less to do with the effectiveness of exercise than the difficulty researchers have in quantifying true creativity.

# Part III

# WHY EXERCISE?

People exercise for so many different reasons. Often how we exercise is determined by who we are, young or old, male or female, or what we wish to achieve by exercising. Some people exercise to lose weight, to build a set of amazing abs, to relieve boredom, to spend more time with friends or make new ones or, conversely, to get away from everyone and everything by taking a long walk alone. Except for losing weight, exercise easily achieves all of these goals. Exercise makes our bones denser and muscles stronger while it improves vascular flow and tissue oxygenation. No matter who you are or what your expectations might be, you will feel better after moderate exercising.

The purpose of this book was to investigate whether your brain, specifically, benefits from exercise. Like most people, I initially assumed that the answer would be an unambiguous "Yes!" I was surprised to discover that the answer was anything but unambiguous. After reviewing the available scientific literature, I came to the conclusion that your brain *does* care whether you exercise, just not always for the reasons that you might think. What I discovered is that your brain benefits the most when you perform activities that it evolved to perform—to move around your environment with purpose, not for diversion or sport. Your brain benefits when the movement addresses its unique evolutionary priorities. Your brain wants you to move around and explore your environment as often as possible so that you can discover the two most important things that are out there to discover; the only two things it evolved to care about: food and sex. Yes, I know that humans find pleasure in things other than food and sex, such as listening to music, appreciating a beautiful sunset, etc. (I will leave the rest of the list to the poets). What I find fascinating is that the pleasure you experience when doing pleasurable things, such as exercising, is entirely related to the activation of the same brain systems and neurotransmitters that are activated when you enjoy food and sex.

Personal survival and the continuation of your species are what brains evolved to achieve for you, as well as your cat, the bird your cat just ate, and every animal that you have ever seen in a zoo. Your brain needs you to move so that it can survive another day in order to find another, hopefully much-better-looking and smarter (if you're lucky), brain with which to propagate the next generation of moving brains. In return, your brain rewards you, if only very briefly, with a euphoria. Movement requires food. The utilization of food for movement is why things get very complicated with regard to answering the question of whether your brain benefits from exercise. For our most ancient ancestors the not-so-simple task of obtaining food could be hazardous to their health if their food source happened to possess claws or sharp teeth. Today, obtaining food is much easier; however, the biochemical processes that are necessary to extract energy from food, due to the demand for a constant supply of oxygen in order to achieve this goal, are hazardous to the health of every one of your cells. This danger exists thanks to one fateful day 1.8 billion years ago when an oxygen-eating simple cell, today called a mitochondrion, was engulfed by a somewhat larger single-celled organism and the two developed a remarkable and wondrous symbiotic relationship. The oxygen-eating simple cell, now safely nestled inside the bigger cell, provided lots of energy that enhanced their joint survival and, ultimately, an additional feature that greatly improved their joint survival (I'm ignoring plants who also contain mitocondria)—movement. Thanks to this symbiosis, life on this planet would never be the same boring thing that had existed for so many billions of years: Sex and death had now arrived!

Thus, your brain greatly benefits from daily movement. But what about exercise? Earlier I defined exercise as a planned, socially structured, and repeated behavior performed with the intent to maintain or improve some aspect of physical fitness. Does exercise serve this purpose effectively? With regard to physical fitness, as I mentioned earlier, your cardiovascular system, no matter how old you are, probably benefits the most from regular aerobic exercise, while your bones benefit from weight-bearing exercises.

However, these benefits come with a cost. You must eat to provide the energy to exercise; obtaining this energy requires the utilization of oxygen. Thus, determining whether exercise benefits your body and brain depends on finding a balance between your body's requirement for oxygen and your ability to defend yourself from it. Moderate exercise adapts your tissues to the consequences of the increased oxidative metabolism required to support movement, while too much exercise, particularly over time, pushes metabolic demands to the point that your cells' innate protective systems are overwhelmed; the increasing levels of toxic ROS lead to tissue injury and accelerated aging. For the brain, the abrupt increase in ROS levels induced by extreme levels of exercising is much more harmful due to its own very high metabolic rate. This is partially why extreme exercising is so harmful to the brain. Thus, moderate exercise is less harmful to the brain because it produces lower levels of harmful ROS molecules that the body and brain are capable of neutralizing.

What is moderate exercise? Obviously, the answer will differ somewhat depending on your age, sex, and health status. When I asked my colleagues what they considered moderate exercise I received the following list: brisk walking, a gentle swim, mowing the lawn, bicycling at a slow pace, tennis (doubles, not singles), ballroom dancing, holding most yoga poses, gardening, or almost any activity that you can maintain for one to two hours and that requires some effort but not so much that you cannot talk while doing it. Most of these exercises are shared experiences that are enjoyable and thus more likely to be included in your weekly schedule.

Moderate exercise also offers many benefits via the production and release of a complex blend of molecules from muscles that modestly improve a range of brain functions. Daily moderate exercise may also delay the functional decay associated with normal aging and many common neurodegenerative diseases that depend on the long-term presence of brain inflammation. The modest improvements in mood or learning and memory abilities in response to daily moderate levels of exercise are most likely related to the reduction in inflammation and augmentation of the blood flow following the influx of VEGF into the brain.

Overall, your muscles behave as though they are a complex endocrine system, much like your thyroid gland or pancreas. Muscles send out chemical signals so that the brain and body are aware of what they are doing. One of the most studied, and over-hyped, of these hormones is BDNF. The reality of BDNF turned out to be much more complicated than I expected. When it comes to understanding the role of BDNF, the devil was most definitely hidden deep within the details. BDNF influences bodily functions that are indirectly related to exercising, such as regulating energy homeostasis and modulating glucose metabolism in peripheral tissues. Via these actions, BDNF mediates the beneficial effects of moderate exercise by stimulating glucose transport into muscle cells and their numerous mitochondria. This vital role might explain why exercising is associated with such high levels of BDNF in the blood and muscles. Science is still without a viable mechanism to explain why blood platelets release so much BDNF following aerobic exercise. Whatever the reason, it probably has nothing to do with the brain. Most recent studies have *failed* to find a valid mechanist link between the increase in blood levels of BDNF and learning and memory abilities.

What about inside the brain? The benefits of aerobic exercise on increased BDNF production inside the brain and adult neurogenesis probably depend on the entry of beta-hydroxybutyrate into the brain from the periphery. Overall, the brain is informed about the activity of muscles by the entry of a subset of hormones, including beta-hydroxybutyrate, VEGF, irisin, and PGC-1alpha, during and after exercise that may significantly benefit brain health and cognitive function. Are the extra BDNF molecules that are produced in the brain responsible for the cognitive benefits associated with exercise? No one knows, yet. Exactly how BDNF acts to improve cognitive function remains a mystery. However, neuroscientists have discovered some potential mechanisms. One possible explanation is that BDNF, working together with VEGF, increases the volume of a region of the hippocampus that is critical for learning and memory; this region is called the *dentate gyrus*. The increase in volume of the dentate gyrus is most likely due to the growth of new blood vessels and not the birth of new neurons—most of your newborn neurons probably do not survive very long. The improvements in learning and memory abilities in response to regular moderate exercise are today believed to be due to the augmentation of the blood flow to regions of the brain that are responsible for these abilities. This makes intuitive sense given that regional changes in blood flow are directly related to which brain regions are being utilized at any point in time. Indeed, this assumption about the intimate relationship between blood flow, the brain's constant requirement for an adequate oxygen supply, and regional brain activity is the basis of the blood-oxygen-level-dependent (BOLD) imaging method used in functional MRI to monitor the different areas of the brain that are active at any given time. The increased blood flow also reduces the level of inflammation due to ROS formation. Essentially, the benefits of exercise on the brain require a delicate dance between various molecules so that the benefits of exercise outweigh the risks.

Exercise is not a panacea. With regard to depression, exercise is not better than drugs; however, daily moderate exercise should be considered as an important adjunct therapy to any treatment plan for depressed or anxious adult patients. My best advice is to do both: Get regular exercise and take your medications. With regard to the aged or injured brain, regular exercising, if the person is able, will offer modest, but usually reliable, cognitive benefits. The main caveat for most old adults is that the greater the degree of cognitive decline a person exhibits prior to beginning any exercise routine, the less likely is the person to benefit from exercise. Many have claimed that regular exercise may slow the onset of Alzheimer's disease. I have spent my entire scientific career studying the neurobiological mechanisms that underlie Alzheimer's disease. In my opinion, this claim is correlational and does not represent a true cause-and-effect connection; on balance, people who exercise tend to be healthier overall and demonstrate fewer poor lifestyle habits. Alzheimer's disease is clearly related to lifestyle factors such as being obese or having diabetes; both of these conditions are related to having lived an unhealthy lifestyle. In general, if you have symptoms of dementia it is probably too late to expect much benefit from any exercise routine.

Throughout the preceding chapters I have emphasized the need to pay attention to causality—what came first and what is the true relationship of factors, such as exercise and any change in brain function. Scientists, politicians, the media, and virtually everyone else are always vulnerable to protopathic bias, which is the act of mistaking what came first in the order of causation. Due to a general lack of knowledge, the scientific and popular media exhibit considerable evidence of protopathic bias with regard to the influence of exercise on brain function.

For most people, under most conditions of good general health, the maximal benefits of exercise will be achieved at moderate levels of daily physical activity. Ironically, similar advice concluded my book *Your Brain on Food*: Too much or too little of any essential drug or nutrient is not healthy. The best advice to achieve good brain and body health has not changed since it was first written down three millennia ago by Homer in *The Odyssey*: "In all things, moderation." Move a little every day and eat a little less than you do now . . . and if all of that fails, try liposuction—it worked for the mice!

# FURTHER READING

Aadland KN et al. (2019) Effects of the Active Smarter Kids (ASK) physical activity school-based intervention on executive functions: A cluster-randomized controlled trial. *Scand J Edu Res* 63:214.

Alkadhi KA (2018) Exercise as a positive modulator of brain function. *Mol Neurobiol* 55:3112.

Amin FM et al. (2018) The association between migraine and physical exercise. *J Headache Pain* 19:83.

Bao D et al. (2019) The effects of fatiguing aerobic exercise on the cerebral blood flow and oxygen extraction in the brain: A piloting neuroimaging study. *Front Neurol* 10:654.

Bardou I et al. (2013) Differential effects of duration and age on the consequences of neuroinflammation in the hippocampus. *Neurobiol Aging* 34:2293.

Bernardo TC et al. (2016) Physical exercise and brain mitochondrial fitness: The possible role against Alzheimer's disease. *Brain Pathol* 26:648.

Birch AM et al. (2019) Lifelong environmental enrichment in the absence of exercise protects the brain from age-related cognitive decline. *Neuropharmacology* 145:59.

Blain B et al. (2019) Neuro-computational impact of physical training overload on economic decision-making. *Cell Biol* 29:1.

Blair SN et al. (1989) Physical fitness and all-cause mortality: A prospective study of healthy men and women. *JAMA* 262:2395.

Chaddock-Heyman L et al. (2019) Physical activity increases white matter microstructure in children. *Front Neurosci* 12:950.

Chen F-T et al. (2019) Effects of exercise modes on neural processing of working memory in late middle-aged adults: An fMRI study. *Front Aging Neurosci* 11:224.

Chew C, Sengelaub DR (2020) Exercise promotes recovery after motoneuron injury via hormonal mechanisms. *Neural Regen Res* 15:1373.

Cobianchi S et al. (2017) Neuroprotective effects of exercise treatments after injury: The dual role of neurotrophic factors. *Curr Neuropharmacol* 15:495.

Conde C et al. (2020) The protective effect of extra-virgin olive oil. *Nutr Neurosci* 23:45.

Cope EC et al. (2018) Microglia play an active role in obesity-associated cognitive decline. *J Neurosci* 38:8889.

Delezie J, Handschin C (2018) Endocrine crosstalk between skeletal muscle and the brain. *Front Neurol* 9:698.

Ehlers DK et al. (2017) The effects of physical activity and fatigue on cognitive performance in breast cancer survivors. *Breast Cancer Res Treat* 165:699.

Fattoretti P et al. (2019) Testosterone administration increases synaptic density in the gyrus dentatus of old mice independently of physical exercise. *Exp Gerontol* 125:110664.

Feliciano DP et al. (2014) Nociceptive sensitization and BDNF up-regulation in a rat model of traumatic brain injury. *Neurosci Lett* 583:55.

Feter N et al. (2018) Effect of physical exercise on hippocampal volume in adults: Systematic review and meta-analysis. *Science & Sports* 33:327.

Freitasa DA et al. (2018) High-intensity interval training modulates hippocampal oxidative stress, BDNF and inflammatory mediators in rats. *Physiol Behav* 184:6.

Fussa J (2015) A runner's high depends on cannabinoid receptors in mice. *Proc Nat Acad Sci USA* 112:13105.

Geisler M et al. (2018) Expectation of exercise in trained athletes results in a reduction of central processing to nociceptive stimulation. *Behav Brain Res* 356:314.

Gholamnezhad Z et al. (2019) Effect of different loads of exercise and *Nigella sativa* L. seed extract on serologic and hematologic parameters in rat. *Indian J Exp Biol* 57:21.

Giacobbo BL et al. (2019) Brain-derived neurotrophic factor in brain disorders: Focus on neuroinflammation. *Molec Neurobiol* 56:3295.

Gomes-Osman J et al. (2018) Exercise for cognitive brain health in aging. A systematic review for an evaluation of dose. *Neurology: Clinical Practice 8*:257.

Harkin A (2014) Muscling in on depression. *N Engl J Med 371*:24.

Holgado D et al. (2019) The effects of transcranial direct current stimulation on objective and subjective indexes of exercise performance: A systematic review and meta-analysis. *Brain Stim 12*:242.

Juarez EJ, Samanez-Larkin GR (2019) Exercise, dopamine, and cognition in older age. *Trends Cogn Sci 23*:986.

Karssemeijer E et al. (2019) The quest for synergy between physical exercise and cognitive stimulation via exergaming in people with dementia: A randomized controlled trial. *Alzheimers Res Ther 11*:3.

Krukowski K et al. (2018) Female mice are protected from space radiation-induced maladaptive responses. *Brain Behav Immun 74*:106.

Kuhn G et al. (2018) Adult hippocampal neurogenesis: A coming-of-age story. *J Neurosci 38*:10401.

Lane N (2005) *Power, sex, suicide: Mitochondria and the meaning of life.* Oxford, UK: Oxford University Press.

Latorre R et al. (2017) Creativity and physical fitness in primary school-aged children. *Pediatr Internat 59*:1194.

Lee HM et al. (2018) Antidepressant drug paroxetine blocks the open pore of Kv3.1 potassium channel. *Korean J Physiol Pharmacol 22*:71.

Lucke-Wold BP (2018) Supplements, nutrition, and alternative therapies for the treatment of traumatic brain injury. *Nutr Neurosci 21*:79.

Maass A et al. (2016) Relationships of peripheral IGF-1, VEGF and BDNF levels to exercise-related changes in memory, hippocampal perfusion and volumes in older adults. *NeuroImage 131*:142.

Maekawa T et al. (2018) Electrically evoked local muscle contractions cause an increase in hippocampal BDNF. *Appl Physiol Nutr Metab 43*:491.

Marchalant Y, Baranger K, Wenk GL, Khrestchatisky M, Rivera S (2012) Can the benefits of cannabinoid receptor stimulation on neuroinflammation, neurogenesis and memory during normal aging be useful in AD prevention? *J Neuroinflamm 9*:10.

Marchalant Y, Brothers HM, Norman GH, Karelina K, DeVries AC, Wenk GL (2009) Cannabinoids attenuate the effects of aging on neuroinflammation and neurogenesis. *Neurobiol Dis 34*:300.

Marchalant Y, Brothers HM, Wenk GL (2008) Inflammation and aging: can endocannabinoids help? *Biomed Pharmacother* 62:212.

Marchalant Y, Brothers HM, Wenk GL (2009) Cannabinoid agonist WIN-55,212-2 partially restores neurogenesis in the aged rat brain. *Molec Psychiat* 14:1068.

Marchalant Y, Cerbai F, Brothers HM, Wenk GL (2008) Cannabinoid receptor stimulation is anti-inflammatory and improves memory in old rats. *Neurobiol Aging* 29:1894.

Marosi K, Mattson MP (2014) BDNF mediates adaptive brain and body responses to energetic challenges. *Trend Endocrinol Metab* 25:89.

Mattson MP (2019) An evolutionary perspective on why food overconsumption impairs cognition. *Trends Cogn Sci* 23:200.

Mitre M et al. (2017) Neurotrophin signalling: novel insights into mechanisms and pathophysiology. *Clin Sci (Lond)* 131:13.

Ogonovszky H et al. (2005) The effects of moderate-, strenuous- and over-training on oxidative stress markers, DNA repair, and memory, in rat brain. *Neurochem Int* 46:635.

O'Leary JD et al. (2019) Differential effects of adolescent and adult-initiated voluntary exercise on context and cued fear conditioning. *Neuropharmacology* 145:49.

Oppezzo M, Schwartz DL (2014) Give your ideas some legs: The positive effect of walking on creative thinking. *J Exp Psychol Learn Mem Cogn* 40:1142.

Panab W et al. (1998) Transport of brain-derived neurotrophic factor across the blood–brain barrier. *Neuropharmacology* 37:1553

Park S-S et al. (2019) Exercise attenuates maternal separation-induced mood disorder-like behaviors by enhancing mitochondrial functions and neuroplasticity in the dorsal raphe. *Behav Brain Res* 372:112.

Pengfei K (2020) Relationship between exercise and 18-fluorodeoxyglucose positron emission tomography/computed tomography imaging of brain in the elderly. *J Med Imag Hlth Informat* 10:1962.

Pontzer H (2018) Energy constraint as a novel mechanism linking exercise and health. *Physiology* 33:384.

Qingfeng X et al. (2019) Treadmill exercise ameliorates focal cerebral ischemia/reperfusion-induced neurological deficit by promoting dendritic modification and synaptic plasticity via upregulating caveolin-1/VEGF signaling pathways. *Exp Neurol* 313:60.

Radak T et al. (2016) Physical exercise, reactive oxygen species and neuroprotection. *Free Radic Biol Med 98*:187.

Roeh A et al. (2019) Depression in somatic disorders: Is there a beneficial effect of dxercise? *Front Psychiatry 10*:141.

Rosi S et al. (2009) Accuracy of hippocampal network activity is disrupted by neuroinflammation: Rescue by memantine. *Brain 132*:2464.

Serra-Millàs M (2016) Are the changes in the peripheral brain-derived neurotrophic factor levels due to platelet activation? *World J Psychiatr 6*:84.

Shaif NA et al. (2018) The antidepressant-like effect of lactate in an animal model of menopausal depression. *Biomedicines 6*:108.

Sharifi M et al. (2018) The effect of an exhaustive aerobic, anaerobic and resistance exercise on serotonin, beta-endorphin and BDNF in students. *Phys Ed Stud 22*:272.

Shishmanova-Doseva M et al. (2019) Enhancing effect of aerobic training on learning and memory performance in rats after long-term treatment with Lacosamide via BDNF-TrkB signaling pathway. *Behav Brain Res 41*:370.

Singhal G et al. (2019) Short-term environmental enrichment, and not physical exercise, alleviate cognitive decline and anxiety from middle age onwards without affecting hippocampal gene expression. *Cogn Affect Behav Neurosci 19*:1143.

Sleiman SF et al. (2016) Exercise promotes the expression of brain derived neurotrophic factor (BDNF) through the action of the ketone body beta-hydroxybutyrate. *ELIFE 5*:e15092.

Smith PA (2014) BDNF: No gain without pain? *Neuroscience 283*:107.

Son JS et al. (2018) Exercise-induced myokines: A brief review of controversial issues of this decade. *Expert Rev Endocrinol Metab 13*:51.

Spindler C et al. (2019) Paternal physical exercise modulates global DNA methylation status in the hippocampus of male rat offspring. *Neural Regen Res 14*:491.

Stein AM (2018) Physical exercise, IGF-1 and cognition: A systematic review of experimental studies in the elderly. *Dement Neuropsychol 12*:114.

Teixeira A et al. (2009) Intense exercise potentiates oxidative stress in striatum of reserpine-treated animals. *Pharm Biochem Behav 92*:231.

Tsai S-J (2018) Critical issues in BDNF Val66Met genetic studies of neuropsychiatric disorders. *Front Mol Neurosci 11*:156.

Vilela TC et al. (2018) Aerobic and strength training induce changes in oxidative stress parameters and elicit modifications of various cellular components in skeletal muscle of aged rats. *Exp Gerontol* 106:21.

Wang JJ et al. (2020) Moderate exercise has beneficial effects on mouse ischemic stroke by enhancing the functions of circulating endothelial progenitor cell-derived exosomes. *Exp Neurol* 330:113325.

Wenk GL (1989) Nutrition—cognition and memory. In RB Weg (Ed.), *Topics in geriatric rehabilitation, Volume 6: Nutrition and rehabilitation.* Rockville, MD: Aspen Publishers, p. 79.

Wenk GL (1992) Dietary factors that influence the neural substrates of memory. In RL Isaacson, K Jensen (Eds.), *The vulnerable brain and environmental risks. Volume 1: Malnutrition and hazard assessment.* New York, NY: Plenum, p. 67.

Wenk GL (1999) Functional neuroanatomy of learning and memory. In DS Charney, EJ Nestler, BS Bunney (Eds.), *Neurobiology of mental illness* (1st ed.). New York, NY: Oxford University Press, p. 679.

Wenk GL (2000) Drugs and intelligence. In AE Kazdin (Ed.), *Encyclopedia of psychology. Volume 3.* Washington, DC: American Psychological Association & New York, NY: Oxford University Press, p. 101.

Wenk GL (2007) Neurodegenerative diseases and memory: A treatment approach. In RP Kesner, JL Martinez (Eds.), *Neurobiology of learning and nemory* (2nd ed.) Elsevier, p. 519.

Wenk GL (2017) *The brain: What everyone needs to know.* Oxford, UK: Oxford University Press.

Wenk GL (2019) *Your brain on food: How chemicals control your thoughts and feelings* (3rd ed.). Oxford, UK: Oxford University Press.

Wenk GL, Olton DS (1989) Cognitive enhancers: Potential strategies and experimental results. *Prog Neuropsychopharmacol Biol Psychiat* 13:S117.

Wollseiffen P et al. (2018) Distraction versus intensity: The importance of exercise classes for cognitive performance in school. *Med Princ Pract* 27:61.

Won J et al. (2019) Brain activation during executive control after acute exercise in older adults. *Inter J Psychophys* 146:240.

Yau SY et al. (2014) Physical exercise-induced hippocampal neurogenesis and antidepressant effects are mediated by the adipocyte hormone adiponectin. *Proc Nat Acad Sci USA* 111:15810.

Zahl T et al. (2017) Physical activity, sedentary behavior, and symptoms of major depression in middle childhood. *Pediatrics 139*:1.

Zhang Y, Pardridge WM (2001) Conjugation of brain-derived neurotrophic factor to a blood–brain barrier drug targeting system enables neuroprotection in regional brain ischemia following intravenous injection of the neurotrophin. *Brain Res 889*:49.

Zhang Y, Pardridge WM (2001) Neuroprotection in transient focal brain ischemia after delayed intravenous administration of brain-derived neurotrophic factor conjugated to a blood–brain barrier drug targeting system. *Stroke 32*:1378.

Zsuga J et al. (2018) Blind spot for sedentarism: Redefining the diseasome of physical inactivity in view of circadian system and the irisin/BDNF axis. *Front Neurol 9*:818.

# INDEX

For the benefit of digital users, indexed terms that span two pages (e.g., 52–53) may, on occasion, appear on only one of those pages